Charles Cardale Babington

The British rubi

An attempt to discriminate the species of Rubus known to inhabit the British Isles

Charles Cardale Babington

The British rubi
An attempt to discriminate the species of Rubus known to inhabit the British Isles

ISBN/EAN: 9783337224530

Printed in Europe, USA, Canada, Australia, Japan

Cover: Foto ©Andreas Hilbeck / pixelio.de

More available books at **www.hansebooks.com**

THE BRITISH RUBI.

𝕮𝖆𝖒𝖇𝖗𝖎𝖉𝖌𝖊:

PRINTED BY C. J. CLAY, M.A.
AT THE UNIVERSITY PRESS.

THE

BRITISH RUBI:

AN ATTEMPT TO

DISCRIMINATE THE SPECIES OF RUBUS KNOWN TO INHABIT THE BRITISH ISLES.

BY

CHARLES CARDALE BABINGTON, M.A. F.R.S.

PROFESSOR OF BOTANY IN THE UNIVERSITY OF CAMBRIDGE.

London:

JOHN VAN VOORST, PATERNOSTER ROW.

1869.

NOTICE.

IT was intended that a series of Quarto Plates should accompany this Essay. But as after much unavoidable delay it is still impossible for them to be very soon ready for publication, the Author has thought it unadvisable to defer the issue of the description and remarks upon the species. He hopes on a future occasion to publish the plates as a distinct work.

The AUTHOR desires to express his grateful acknowledgments to the SYNDICS OF THE UNIVERSITY PRESS for their liberality in granting him the expense of the Printing and Paper of this Volume.

Species Ruborum [Britannicas] diligenter examinare et descriptionibus, quoad fieri potuerit, perfectis illustrare conati sumus, memores verborum Linnæi " ne varietas loco speciei sumatur, ubique cavendum est." ARRH. *Monog.* 58.

PREFACE.

———◆———

HAVING acquired, chiefly through the kindness of different botanists, but partly by purchase, what is probably by far the most extensive collection of Brambles which has ever been formed in this country, I have thought it well to draw up such an account of them as these opportunities, and a tolerably long-continued study of *Rubi*, has enabled me to prepare. My collection contains nearly all Leighton's *Rubi*, including the specimens submitted by him to the examination of Nees von Esenbech, Borrer, and Lindley, and named and commented upon by them ; a very complete set of Bloxam's specimens, and also of those of Bell Salter and Lees. I have many specimens named by Borrer, Coleman, Hort, and other students of this difficult genus ; have myself collected Brambles extensively in various parts of the kingdom, and have had many species long in cultivation in the Cambridge Botanic Garden. This account of the collection will shew that the opportunities within my reach are such as to render it probable that at least some valuable results may be attained by its study. Unfortunately there is great diffi-culty in obtaining authentic specimens of such unwieldy

a 3

plants from continental botanists, nevertheless my Herbarium
does contain a considerable number of them. Reichenbach's
Flora exsiccata supplies a few, Fries's invaluable *Herbàrium
normale* others, Wirtgen's *Herbarium Ruborum Rhenanorum*
a considerable number; the Abbe Questier has given to me
a very extensive series of the French species, and I possess
others obtained from Dr F. Schultz and Professor C. Billot.
Unfortunately no botanist in Britain is known to have
typical specimens of the plants figured in the *Rubi Germanici*, of those described by Godron in the *Flore de France*,
or by Boreau in his *Flore du centre de la France*. A few
also of those included in Arrhenius's valuable *Monographia
Ruborum Sueciæ* are unknown to me, although several which
I did not before possess have been kindly sent by Mr Joh.
Lange of Copenhagen.

It is believed that the following essay will afford the
means of determining many, perhaps most, of our species :
but it is only by careful and long-continued study that any
person can expect to attain a correct knowledge of such
difficult plants. My hope is that the readers of this book
will endeavour to correct the mistakes into which I am
sure to have fallen ; for with the utmost care, and I may
venture to add that no care has been neglected, I cannot
avoid feeling convinced that the truth has only been
approached, and that perhaps rather distantly in some
cases.

Several botanists of the highest eminence both in this
country and upon the European continent have thought that
all our brambles are infinitely varying hybrids or forms of

one, two, or as some think four species. Such an opinion
is the natural result of an examination of a few specimens,
perhaps not very perfect, preserved in an Herbarium. But
if much study of the plants in their native places of growth
is combined with that of an extensive series of preserved
specimens of each form, it does seem to me that nature possesses many more species than those distinguished men are
prepared to admit. It is quite likely that the time may
come when several of the forms here looked upon as species
will be shewn not to be distinct from others. The many
blunders which have been made by myself and other students
of this difficult genus would make it very presumptuous in
me to think otherwise. Dr Godron in his valuable essay *Le
genre Rubus considéré au point de vue de l'espèce, (Mém. de
la Société R. de Nancy*, 1849, p. 210), has shewn con-
clusively that we cannot reduce all the European *Rubi* to
the species described by Linnæus, and proved that there are
real and constant characters by which to distinguish as true
species many more than were known to the great Swedish
botanist.

As a contrast to the idea that the number of species is
small, it may be mentioned that a German botanist, (P. J.
Müller) has published (16 and 17 *Jahresberichte der Pollichia*) descriptions of what are supposed by him to be 236
species of *Rubi* inhabiting France and Germany. I believe
these to be descriptions of mere forms. They might have
been of considerable value had their author more frequently
identified his plants with those of other botanists, or pointed
out their distinctions from them : also, if he had given

specific characters or even a synoptical table of his plants.
Without either of these helps it is exceedingly difficult to
identify them.   Fortunately specimens of a considerable
number of them are published in Wirtgen's excellent *Herb.*
Ruborum Rhenanorum, and many others are contained in
Mr J. G. Baker's extensive Herbarium, which he most
kindly placed for a lengthened period in my hands.

CAMBRIDGE,
May 1, 1869.

GEOGRAPHICAL DISTRIBUTION OF THE SPECIES.

Rubus.	Peninsula.	Channel.	Thames.	Ouse.	Severn.	S. Wales.	N. Wales.	Trent.	Mersey.	Humber.	Tyne.	Lakes.	W. Lowlands.	E. Lowlands.	E. Highlands.	W. Highlands.	N. Highlands.	N. Isles.	S. Atlantic.	Blackwater.	Barrow.	Leinster Coast.	Liffey and Boyne.	Lower Shannon.	Upper Shannon.	N. Atlantic.	N. Connaught.	Erne.	Donegal.	Ulster Coast.
Provinces.	1	2	3	4	5	6	7	8	9	10	11	12	13	14	15	16	17	18	19	20	21	22	23	24	25	26	27	28	29	30
1. Idaeus	1	2	3		5	6	7	8	9	10	11	12	13	14	15	16	17	18	19	20	21	22	23	24	25	26	27	28	29	30
2. Leesii	1		3		5	6	7	8	9	10	11	12	13	14	15	16			19											30
3. suberectus					5	6	7	8	9	10	11	12	13	14	15	16			19						25					30
4. fissus		2	3		5	6	7	8	9	10	11	12	13	14	15	16			19					24		26				30
5. plicatus		2	3		5	6	7	8	9	10	11	12	13	14	15	16			19				23		25					30
6. affinis	1	2	3		5	6	7	8	9	10	11	12	13	14	15	16			19							26				30
7. Lindleianus	1	2	3	4	5	6	7	8	9	10	11	12	13	14	15								23	24						
8. rhamnifolius	1	2	3	4	5	6	7	8	9	10	11	12	13	14	15	16														
9. incurvatus	1											12	13	14	15	16														
10. imbricatus		2	3	4	5		7	8	9	10	11	12	13																	
11. latifolius	1	2			5	6	7	8	9	10	11	12									21									
12. discolor	1	2	3	4	5	6	7	8	9	10	11	12	13	14	15	16			19		21		23	24				28		30
13. thyrsoideus	1	2	3	4	5	6	7	8	9	10	11	12	13	14	15	16			19				23							30
14. leucostachys	1	2	3		5	6	7	8	9	10	11	12		14					19					24		26				30
15. Grabowskii	1	2			5	6	7	8		10																				
16. Colemanni	1							8		10																				
17. Salteri	1	2	3		5	6	7	8	9		11	12							19											30
18. carpinifolius		2	3		5	6	7	8	9	10	11	12				16			19											

	1	2	3	4	5	6	7	8	9	10	11	12	13	14	15	16	17	18	19	20	21	22	23	24	25	26	27	28	29	30
19. villicaulis	1	2	3		5		7	8		10	11	12		14	15		17		19	20			23			26				30
20. macrophyllus	1	2	3	4	5	6	7	8	9	10	11	12	13	14	15	16			19				23							30
21. mucronulatus	1	2	3		5		7	8		10	11			14	15	16														
22. Sprengelii		2	3		5			8	9	10		12							19											
23. Bloxamii	1		3	4	5		7	8	9	10	11								19						25					
24. Hystrix	1	2	3	4	5	6	7	8	9	10	11	12	13	14					19											
25. rosaceus	1	2	3		5		7	8	9	10	11	12																		
26. pygmaeus	1	2	3		5		7	8		10	11	12	13	14																
27. scaber	1	2	3		5	6	7	8		10	11	12	13	14	15							22	23							
28. rudis	1	2	3	4	5	6	7	8	9	10	11	12	13	14								22	23							
29. Radula	1	2	3	4	5		7	8	9	10	11	12	13	14	15	16			19											30
30. Koehleri	1	2	3	4	5		7	8	9	10	11	12	13	14	15															
31. fusco-ater		2	3		5		7	8		10	11																			
32. diversifolius	1	2	3	4	5	6	7		9	10		12																		
33. Lejeunii		2	3		5		7	8	9	10	11	12																		
34. pyramidalis	1		3		5		7			10	11	12	13		15															
35. Güntheri	1		3		5		7	8	9	10	11	12	13																	
36. humifusus	1		3		5		7		9	10	11	12	13		15															
37. foliosus	1		3		5		7	8	9	10	11	12	13						19	20										30
38. glandulosus		2	3	4	5	6	7		9	10		12		14					19		21									
39. Balfourianus	1	2	3	4	5	6	7	8	9	10		12	13		15	16				20										30
40. corylifolius	1	2	3	4	5		7	8	9	10	11	12	13													26				
41. altheifolius	1	2	3	4	5	6	7	8	9	10	11	12									21									
42. tuberculatus		2	3	4	5	6	7		9	10	11	12	13	14					19	20			23							
43. caesius	1	2	3	4	5	6	7	8	9	10	11		13	14	15				19	20			23	24		26	27	28		30
44. saxatilis		2	3		5		7	8	9	10	11	12		14	15	16		18	19					24				28	29	30
45. Chamaemorus										10	11				15													28		

CONTENTS.

HISTORICAL SKETCH.

———◆———

A SHORT account of the progress made in the study of the fruticose Brambles by English botanists will probably possess some interest. We may commence with our great naturalist Ray. In his earliest work, *Catalogus Plantarum circa Cantabrigiam nascentium* (1660), he records two species (1) *R. minor fructu cæruleo* [*R. cæsius* Linn.], and (2) *Rubus* [*R. discolor* W. and N.]; in his *Catalogus Plantarum Angliæ* (1670) four are recorded, viz. the same two, and *R. Idæus*, and *R. alpinus humilis* [*R. saxatilis* Linn.]; in the *Synopsis Methodica* (1690) he separates *R. saxatilis* from the other species because of its being herbaceous, placing it in the same group with *R. Chamæmorus*, thus leaving three fruticose species. To these he added in the 3rd edition of the *Synopsis* (1724) a white-fruited plant found near Oxford by Bobart, which cannot have been more than a chance variety of some species, and is not now capable of determination. It may very probably have belonged to *R. thyrsoideus* (*R. fruticosus* W. and N.); for there is a variety of that plant named "*leucocarpus*, carpellis albis" recorded by Seringe in De Candolle's *Prodromus* (ii. 561). He seems therefore not to have distinguished more than three real species. It is

1

curious, as will be seen below, that an eminent botanist, publishing in 1858, returns to the precise view of the subject entertained in 1690 by Ray. Dillenius in his edition of Ray's *Synopsis* (1724) added a plant from Doody's manuscripts, and supposed that there was another allied to No. 2, which I have above supposed to be *R. discolor*. Owing to the short and incomplete descriptions of the earlier botanists it is very difficult to determine their plants.

The next work which is deserving of notice is Hudson's *Flora Anglica*, of which neither edition (1762 and 1778) contains more than the same species, namely, *R. Idæus*, *R. fruticosus* (*R. discolor*) and *R. cæsius:* nor do we find any addition to them before the publication of Smith's *Flora Britannica* (1800), where *R. corylifolius*[1] first appears; for Smith's quotation under it of Withering's *Botanical Arrangement* (ed. 3, 1796) is of very doubtful correctness. It seems to me that Withering, and those who preceded him, had no clear views concerning the plants, and that more than one species (probably several, as we now understand them), were confounded under the name of *R. fruticosus*, and even under its supposed variety *R. fruticosus major*. We should not, however, forget that Mr W. Hall. had published, in 1794, his *R. nessensis* in the *Transactions of the Royal Society of Edinburgh* (Vol. iii.). Smith seems to have been altogether ignorant of this fact; for even in his

---

[1] When Smith proposed to add *R. corylifolius* to the then meagre list of English *Rubi* it was considered as a great innovation. Dalton, a botanist of eminence in his day, wrote to Winch, Oct. 26, 1804, as follows : "I have long been an unbeliever with regard to *Rubus corylifolius*. Brunton says that he knows the plant and believes it a good species. I will talk with him on the subject and procure you a specimen from him. I have it not." Winch's *Correspondence* (*Lin. Soc. Lond.*).

English Flora he quotes Hall's paper as one with which he had no personal acquaintance. In the year 1815 Mr George Anderson gave in the *Linnean Transactions* a full description of what is usually supposed to have been Hall's plant under the far better name of *R. suberectus;* properly taking advantage of the fact that its first publisher furnished a very insufficient account of it, to replace the original name by one which avoids the great objection of being derived from that of a locality of very limited extent. Remarks upon Hall's plant will be found under *R. suberectus* in this book.

Before the appearance of the second volume of Smith's *English Flora* (1824) only a portion of the great work of Weihe and Nees von Esenbech on the German *Rubi* had been published, and Sir James expressed his grief that he was thus prevented from availing himself to a greater extent of the labours of those celebrated botanists. In the *English Flora* Smith describes eleven fruticose species; a great increase from the four recognised in his *Flora Britannica.* These plants are (1) *R. fruticosus,* which we now call *R. discolor;* (2) *R. plicatus;* (3) *R. rhamnifolius,* which includes the *R. cordifolius* of Weihe and Nees; (4) *R. leucostachys;* (5) *R. glandulosus,* now shown to be typically the *R. Koehleri* of Weihe, although it probably included some other glandular brambles; (6) *R. nitidus,* which Borrer states in the *Supplement to English Botany* (fol. 2714) to be the *R. plicatus* of the *Rubi Germanici,* at the same time informing us that "The specimens from Dr Williams, described in the *English Flora* as *R. plicatus,* bear a close resemblance to *R. rhamnifolius,* and probably belong to it;" (7) *R. affinis,* which Borrer and the late Mr Edw. Forster (*Suppl. to Eng. Bot.* f. 2605) unhesitatingly refer to *R. pallidus* of

Weihe; (8) *R. suberectus;* (9) *R. Idæus;* (10) *R. corylifolius;* and (11) *R. cæsius.*

We now turn to Lindley's *Synopsis of the British Flora,* ed. 1 (1829), where there are twenty-four species enumerated and shortly characterized in accordance with the "truly excellent Monograph of the German *Rubi* by Drs Weihe and Nees." I shall not here enter upon a discussion of these plants, for they will be found noticed under the respective species to which they are considered as referrible, but simply state that three supposed new species are recorded, viz. *R. abruptus,* now known to be a state of *R. discolor; R. diversifolius,* concerning which much discussion has arisen, either from some mistake in the naming of specimens or from the displacement of a label in the garden of the Horticultural Society[1]; and *R. echinatus,* a plant apparently ranging under *R. Koehleri.* In the second edition of the *Synopsis* (1835), Lindley quite altered his views concerning brambles; for, although he still gives short characters for eighteen plants, he states that "if it had been possible to prove the four species [*R. suberectus, fruticosus, corylifolius* and *cæsius*] to be themselves physiologically distinct," he would have then "reduced all the others" to them; but as proof even of that seemed to him to be wanting, he adopted a middle course, and grouped the species of his former edition into sections as *Idæi, Suberecti, Corylifolii, Cæsii,* and *Fruticosi.* He also made some alterations in the nomenclature by calling his former *R. fastigiatus = R. fissus,* his *R. echinatus = R. rudis,* his *R. pallidus = R. Koehleri,* and reducing a few of the other

---

[1] If specimens are to be believed this is a distinct species closely allied to *R. fuscoater;* but if the bush in the garden is the authority (although repudiated by Lindley) then it is *R. leucostachys.*

supposed species to the position of synonyms. In what is called a third edition of the *Synopsis* there is no alteration.

A most valuable account of the Brambles, from the pen of Mr Borrer, was published in the second (1831) and third (1835) editions of Hooker's *British Flora.* This must be considered as the groundwork upon which a real knowledge of our native species is founded, and I have derived very great advantage from its study. In the fourth and fifth editions of the *British Flora* only a very short abstract of Mr Borrer's "copious observations" will be found; in the sixth and seventh they are altogether neglected, and a note is inserted stating that the authors (for then Dr Walker-Arnott was associated with Sir W. J. Hooker in the authorship of the book) are "Almost quite convinced …that the characters…are not permanent," and that the reputed species are not "physiologically distinct, all passing into each other without any fixed assignable limit;" and, "from a consideration of what is requisite to constitute a difference between the other Europæan species of *Rubus,* that all of the present section [the *Fruticosi*] are mere varieties approaching on the one side the *R. Idæus,* on the other to *R. saxatilis,* with both of which many fertile and permanent hybrids may have been formed, and are still forming" (*Brit. Fl.* ed. 6, p. 120; ed. 7, p. 122). This view had previously been carried out to its legitimate conclusion by Spenner in his *Flora Friburgensis* (1829), where, under the name of *R. polymorphus,* all our *Rubi Fruticosi* are combined. Spenner says nothing about hybrids, but places what he believes to be one variable species in the same rank with *R. Idæus* and *R. saxatilis.* It seems probable that this was the view also taken by Messrs Hooker and Ar-

nott: for I cannot suppose that they believed the plants to
be hybrids between *R. Idæus* and *R. saxatilis.* It is admitted by zoologists that hybrids are of exceedingly rare occurrence when animals are left to their natural instincts,
although they are not unfrequent between domesticated
species: also, that it is in the highest degree doubtful if a
really hybrid race exists even in domestication. Is it likely
that less care would be taken to keep the species of plants
free from intermixture than is believed to have been exer-
cised in the animal kingdom? Certainly a few fertile hybrids have been obtained artificially, but all the experiments,
accounts of which have fallen in my way, tend to show
that even if isolated they revert to one or other of the parent species in a few generations. As Fries has more than
once remarked, to affix the stigma of hybridity is a convenient mode of escape from many difficulties, but it is not
therefore the more likely to be just. "Ad hybriditatum
voluptas trahit omnes, (1) qui de specierum limitibus dubii
rem absolutam fingunt: 'videtur hybrida planta;' (2) qui
omnes recentius distinctas species ex arbitrio delere student:
'est hybrida forsan planta;' (3) quibus pravas ut species
tueantur, necesse est manifestos transitus 'pro hybridis for-
mis' declarare. At cum hybridas suas species haud limitare
valent, ulterius hybriditates hybriditatum tertii, c. s. p. gra-
dus urgent" (*Fries Symbolæ ad Historiam Hieraciarum,*
p. xxxii.). In the eighth edition of the *British Flora,* the
authors state their belief that the British fruticose species
"might be advantageously reduced to five...which five
would then accord with the four sections into which Mr
Babington has now divided the group."

In the above-mentioned editions (2nd and 3rd) of
Hooker's *British Flora,* Borrer adds only two to the species

already recognized by Smith, although the names of several are corrected, and the characters very much improved; they are *R. carpinifolius* (which is not that so named by the German authors, but seems to be very closely allied to *R. Grabowskii*), and *R. macrophyllus*.

In 1837 Professor David Don drew up a very concise account of these plants for Dr Macreight's *Manual of British Botany*.

In 1841 Leighton published his *Flora of Shropshire*, in which he endeavoured to determine the plants of Nees von Esenbech, Lindley, and Borrer, by transmitting specimens of the Shropshire Brambles to each of them, and obtaining in return their remarks upon the plants. The results are not as satisfactory as might have been expected; for the opinions received are very contradictory, and appear sometimes, especially in the case of the first-named author, not to accord with the descriptions previously published. In 1848 Leighton made a series of most valuable remarks upon some of the same plants in the third volume of the *Phytologist*. Unfortunately his intention of sending further papers on the subject was not fulfilled.

In 1847 Mr Edwin Lees communicated the specific characters of the species, as known to him, to Steele's *Handbook of Field Botany;* and more recently described many of them in his own *Botany of Malvern*, ed. 2 (1852).

Dr Thomas Bell Salter inserted valuable remarks upon these plants in the *Annals of Natural History* (Ser. 1. xv. and xvi. in 1845), and in the *Phytologist* (ii.). He gave a complete synopsis of his views in the *Botanical Gazette* (ii. in 1850), and repeated it, with little or no alteration, in Hooker and Arnott's *British Flora* (ed. 6. in 1850), and Bromfield's *Flora Vectensis* (1856).

In 1846 the author of this book published in the
Annals of Natural History (Ser. 1. xvii.), and also in a
separate form, a *Synopsis of British Rubi*, adding in the
same journal supplements to it in 1847 (Ser. 1. xix.), 1848
(Ser. 2. ii.) and 1852 (Ser. 2. xx.), and has given the characters of all the species supposed to inhabit Britain in the
successive editions of his *Manual of British Botany* (1843,
47, 51, 56, 62, and 67).

In 1850 the Rev. Andrew Bloxam supplied to Miss
Kirby's *Flora of Leicestershire* a very excellent account of
the species which he had ascertained to grow in that county.

In 1851 the Rev. F. J. A. Hort published a new species
(*R. imbricatus*) in the *Annals of Natural History* (Ser. 2.
vii. 374).

In 1853 Dr George Johnston included several brambles
in his elegant *Botany of the Eastern Borders*, and mentions
the curious fact that the vicar of Norham received tithes of
Blackberries (Rubi majores) in the year 1364 (Raine's *North
Durham*, 278).

In 1858 Mr Bentham published his *Handbook of the
British Flora*, wherein he reduces our *Rubi Fruticosi* to
three, viz. *R. Idæus*, *R. fruticosus*, and *R. cæsius;* thus including all of them, except *R. Idæus* (which comprises *R.
Leesii*), under two species. He states it to be his opinion
that the supposed series, even when thus restricted, will
"very frequently be found to pass imperceptibly into each
other." Had not the plan of his whole work been founded
upon a similar principle, this might have been considered
an easy way of appearing to escape from a difficulty.

Mr Irvine included the *Rubi* in his *Illustrated Handbook* in 1858, and states in the Preface that they are described "in conformity generally with Mr Babington's views."

Unfortunately I must decline being considered as at all answerable for most of the statements there made.

Two species are characterized for the first time in Britain in my *Flora of Cambridgeshire* (1860), viz. *R. althæifolius* and *R. tuberculatus*.

Mr Syme in the "new edition" of *English Botany* has followed my arrangement, calling my species sub-species. Only some of the plants thus ranked as sub-species are represented on the plates, most of those remaining without figures which have not been published in the original *English Botany* or its *Supplement*. Unfortunately the want of attention to the colour and clothing of the leaves which exist in the originals of these plates has not been supplied in this new work.

In 1867 Mr Lees published his *Botany of Worcestershire*, at the end of which he has given his latest views upon the species of *Rubi*. In many respects these accord with my ideas, but in some cases his nomenclature is different, and in others the plants which he had in view are apparently not always the same as mine.

It is believed that this is a tolerably complete account of the progressive study of British Brambles. No attempt is made to treat the writings of continental botanists in a similar manner, but I may name those authors whose works have been of the most use to me in my researches. They are Arrhenius, by his *Monographia Ruborum Suecie* (1840), and his notes inserted in Fries's *Mantissa tertia* (1842) and *Summa Vegetabilium Scandinaviæ* (1846); Bluff and Fingerhuth, in the *Compendium Floræ Germanicæ*, ed. 1 (1825) and ed. 2 (1837); Petermann, in his *Flora Lipsiensis* (1838); Godron, in his *Flore de Lorraine*, ed. 1 (1843), ed. 2 (1857), the *Flore de France* (1848), and his *Monographie des Rubus de Nancy* (1843); and the following works are especially deserving of notice, viz. the *Rubi Germanici* of Weihe and Nees v. Esenbech (1822—27); the *Flore du centre de la France*, by Professor Boreau, edition 3 (1857); a valuable paper by Dr Metsch entitled *Rubi Hennebergenses*, which will be found in the *Linnæa* (xxviii. p. 89), published in 1858, although dated 1856 on the title-page; and Garcke's *Flora von Nord- und Mittel-Deutschland*, ed. 7 (1865).

An interesting history of the study of this genus on the continent will be found in Godron's paper entitled *Le genus Rubus considéré au point de vue de l'espèce*. It is included in the *Mémoires de la Société des Sciences de Nancy* (1849, p. 210), and contains some exceedingly good remarks upon the distinctness of species in opposition to those botanists who with Gmelin and Bentham only recognize the Linnæan species, or with Spenner admit only one fruticose *Rubus* into the flora of Europe. He shows that neither climate, soil, nor variations of light and shade, nor even cultivation, will produce those changes in the form and direction of the stem, the shape and texture of the leaves, the outline and structure of the panicle, the shape of the petals, and the kind of fruit, which are requisite if the theory of the authors just mentioned is adopted. An essay entitled *De l'étude spécifique du genre Rubus* (*Congrès Scientifique de France*, 28 Session, t. iii.) by the Abbe Chaboisseau, is a valuable contribution to the literature of this subject.

I may also refer to the Essays of M. Genevier contained in the *Mém. Soc. Acad. d'Anger* (vols. viii. and x.), and the *Rubi Genevenses* of Dr Mercier attached to Reuter's *Flore de Genève*.

INTRODUCTION.

IN the study of *Rubi* it is requisite to take into consideration the habit of the plant, as well as the form and structure of most of its parts. A want of information in the first of these respects renders it most difficult and often impossible to refer dried specimens to their true species with certainty[1]. All the fruticose species throw out long leafy shoots directly from their roots which do not produce any flowers during the first year. The *Idæi* are sometimes exceptions, for their canes (as the gardeners call them) do sometimes flower at the end in the first autumn. The barren stems, as they are usually called, all rise slightly from the ground at the commencement of their growth, but afterwards take different directions which are characteristic of the several species. They are either (i) *suberect*, that is, nearly upright throughout the greater part of their length but nodding slightly at their slender tops; or (ii) *erect-arcuate*, when they are nearly as erect as in the suberect species, but terminate often in a kind of knot consisting of a number of closely placed leaves and usually numerous prickles, from which in the autumn one or more slender shoots descend directly to the ground, where they take root. The other

---

[1] Caulis in multis plantis ita essenciales præbet differentias, ut eo demto, nulla certitudo speciei. *Linn. Philos. Bot.* § 276, ed. 2, p. 218.

species form an arch of more or less height and extent, but when the shoot again arrives at the ground (if early enough in the season) it is prostrate for some distance, and in the autumn again rises at the end into a very low and small arch so as to present its point directly toward the earth, which it penetrates slightly and takes root. It is convenient to divide these plants into such as are (iii) *arcuate*, that is form a large and lofty arch the end of which often does not reach the earth until late in the autumn, when its point immediately pushes itself into the ground and takes root; and the (iv) *arcuate-prostrate*, whose stems, when unsupported, form a very slight and inconspicuous arch, but lie, through-out the greater part of their length, quite close to the ground, often following all the slight inequalities of its surface. The observation of these differences is rendered difficult by the stems being supported by bushes or even by other parts of the plant itself and not reaching the ground, as they would have done if without support. We often see *R. discolor*, which is an arcuate-prostrate plant, rising out of the tops of lofty hedges, and sometimes rendered unable to reach the ground before its growth is stopped by the winter. When thus circumstanced it lies upon the top of · the hedge in precisely the same manner as it would have done upon the ground if not artificially raised. When so pre-vented from taking its more natural position it frequently forms a knot similar to that of the erect-arcuate plants, and tries by throwing out a slender autumnal shoot to arrive at the earth; or extends its growth from the same point during the succeeding summer, frequently, if the thicket is dense, with a like failure : but where such supports are wanting, its stem will be found to form an arch of only a few inches in height, after which it extends to a great length close to the ground, until the final small autumnal arch is produced by means of which the growing point is enabled to bury itself. On the other hand, its ally *R. thyrsoideus* forms a lofty arch even when totally without other support than its own strength, and generally takes root as soon as the end arrives at the soil, never running far along the surface.

The stem is round or has five bluntish angles, between which the faces, although often furrowed, are usually nearly flat. Sometimes the lower part is round and the upper angular. The colour of the stem, as is well remarked by Arrhenius (p. 9), is variable according to the place where the plant grows. In shade it is green or greenish, in spots where it is fully exposed to the light of the sun it usually becomes more or less red or purple, and often acquires a very dark tint of the latter colour; but some species seem to have a greater tendency to assume the dark tint than is possessed by others. The prickles are uniform in shape and direction throughout the stem; or the lower ones are straight and slender, but the rest much stronger, and either patent (that is, at right angles to the stem) or deflexed or declining (when they are straight, but directed downwards). In some species they are all of nearly equal size and placed chiefly or wholly upon the angles of the stem; in others they are very variable in size and scattered over the faces as well as the angles. In the latter case there is usually a very gradual decrease in their size, so that the smallest prickles are not distinguishable from the slender rigid bristles called *aciculi*. The *aciculi* again decrease in strength, and each becomes tipped with a gland, when they take the name of setæ[1].

---

[1] The term *seta* is usually applied by botanists to a strong bristle, but English writers upon the genera *Rubus*, *Rosa* and *Hieracium* confine

The faces of the stem are often furnished with many nearly sessile glands in some of those species which usually have neither setæ nor aciculi. Some stems are quite hairless; but others are more or less thickly covered with hairs, which are either solitary and patent, or two or three spring from the same spot and diverge so as, if numerous, to interlace with those of the neighbouring clusters. In some cases there is more or less fine down, formed of clustered but very small hairs spreading close to the surface of the stem; this is called *felt, tomentum,* or *stellate-pubescence.* The stem of a few species is covered with a kind of bloom (is *pruinose*), especially when young. The faces of the stem in the groups called . *Suberecti, Rhamnifolii* and *Villicaules,* are usually marked with parallel longitudinal lines and have a dull appearance; but a few plants (*R. Lindleianus* for example) have shining faces to their stems.

The leaves are either *pinnate* with seven leaflets, of which four spring from the same spot in opposite pairs, but the upper three (also seated at one spot) are separated from them by a considerable space; or, the upper three consist of an opposite pair similarly separated from the lower four, but the terminal leaflet is again raised above them by a short stalk; both of these combinations are called *septenate-pinnate* leaves: or, (in *R. Idæus*) five leaflets are arranged in an *impari-pinnate* manner: or, the leaf consists of three or five leaflets, all springing from the same place, of which the lower are stalked or sessile, but the terminal is always stalked; these are the *quinate* leaves: or, the two lower are

it (in describing those plants) to the *longish gland-tipped hairs.* Some confusion is caused by this latter use of the term, but, in the want of a special name for those organs, it is probably better to retain its use than to employ the circumlocution of *gland-tipped hairs;* for descriptions are thereby much shortened and facilitated.

placed severally upon the stalks of the two intermediate leaflets, when the leaf is *pedate*. The *septenate-pinnate* leaf is always distinguishable from that which is truly pinnate by the four lower leaflets being, as is already remarked, inserted at the same spot, and also by the unequal bases of the upper pair, which are very irregularly combined with or separated from the terminal leaflet. Thus the septenate-pinnate leaf is nothing more than an anomalous state of the quinate leaf. Nevertheless it is very rarely found except in three or four of the species. Similarly the *pedate* is perhaps to be considered as a state of the ternate leaf: for it appears to be tolerably certain that the ternate and pedate leaves are interchangeable in the same species, or even individual bramble. The true *quinate* leaf is always digitate, and its leaflets are also always distinct from each other, although the lower or outer pair are sometimes sessile.

The upper surface of the leaflets is usually rather darker in colour than the under side; it is either quite naked, or has a few hairs scattered over it or arranged along the grooves which correspond with the ribs and stronger veins of the under side. The under side is either green, and naked with the exception of more or less dense rows of hairs placed upon the ribs and stronger veins, or even also upon the finer veins; or the surface between the veins, and often the veins themselves, is covered with white or whitish felt (*tomentum*), which is sometimes very fine, but often forms rather a thick and dense coat quite hiding the cuticle. The midribs of the leaflets, and the partial and general petioles, are armed on the under side with prickles taking generally the form of hooks. In describing the leaflets, unless the contrary is expressed, the terminal one alone is noticed; it is usually more or less obovate, often cordate at the base, and frequently acuminate at the tip: but some leaflets are strongly cordate below, and some are abruptly cuspidate. The form, although speaking generally it may be called obovate, is sometimes so much and regularly narrowed below as to become almost wedgeshaped, or it may narrow so slightly as to be very nearly oval, or, in a few cases, the sides are so parallel and the two ends so truncate that the leaflet is almost square, with a central terminal cusp. Many intermediate forms are found to which attention should be paid. In some leaves the lower pair of leaflets partially overlaps the intermediate and, rarely, the latter overlap the terminal leaflet; or, the lower leaflets are directed backwards, toward the petiole, so as to leave a clear space between them and their neighbours. Those differences in the direction of the leaflets are usually constant and therefore deserving of attention; but in some species they are not wholly to be depended upon. It is often very difficult to determine what has been the direction of the leaflets after the specimen has been pressed in preparation for the herbarium. The whole leaf is convex, flat, or concave above, and the edges of the leaflets are either similarly curved or flat; or, the whole leaf may be flat and the edges of the leaflets may curve upwards or downwards so as to be concave or convex. The edges of the concave leaflet are usually wavy. The leaflets are sometimes simply and finely dentate or serrate or doubly so; or the double teeth are so large, especially in the upper half of the leaflet, as to resemble dentate or serrate lobes. These lobes are either directed towards the end of the leaf or their tops turn more or less from it: this seems to be a difference of some value, for there are cases in which individuals belonging to species which usually have well-marked lobed dentition have the lobes reduced to very broad but low double teeth, and then the middle secondary tooth of each of them usually shows a clear tendency to take the forward or the patent direction observable in the typical forms.

The general and partial petioles are flat or furrowed on their upper side and rounded below. Their under sides are also furnished with more or less numerous prickles similar to, but usually rather larger than, those found on the under side of the midrib of the leaflets.

In all the fruticose species the stipules are attached to the petiole at some little distance from its insertion; but herbaceous *Rubi* have their stipules attached to the stem itself. In this respect *R. saxatilis* seems to connect these great divisions, for the stipules of its barren stem are often on the petioles, whilst those of the flowering shoot spring from the stem itself.

The *flowering shoot* grows from buds formed in the axils of the leaves of the barren stem of the preceding year; excepting in some of the *Herbacei*, where the stem is represented by a subterranean creeping rhizome, from which the flowering shoots rise at intervals. Therefore the only difference consists in the fact that the *Fruticosi* have aërial, the *Herbacei* subterraneous stems. The scales which formed this bud are persistent, in a faded condition, at the base of the shoot: they vary in colour, and in their clothing, and may furnish characters of some value when carefully noticed. In the *Idæi* and *Suberecti* the shoots spread in two directions (are distichous), but in the other *Fruticosi* they all turn towards the upper side of the stem. Their leaves are very similar to those of the stem, but much more frequently ternate; the lower are sometimes quinate; and the upper floral leaves are frequently simple.

The panicles are of various forms; their branches are

either racemose or corymbose, and they, as well as the peduncles, spread at different angles. Characters derived from
them are not easily described, and therefore are of less value
to the student than they seem really to be in nature. The
rachis and peduncles, and often the whole of the flowering
shoot, are usually furnished with setæ (even in species the
stems of which have no such organs), often have aciculi,
many hairs, and frequently a thick coat of felt. The setæ
on these parts are sometimes shorter than the hairs (sunken),
and may easily be overlooked when not pointed out by their
peculiar colour. The sepals are usually clothed similarly to
the peduncles; they differ considerably in shape and direction when accompanying the fruit. They either end in a
minute point, or a linear or flattened and leaf-like append-
age. In considering the characters derived from this ap-
pendage, it is its presence, not absence, that is supposed to
be of value. For those plants which usually possess the leaflike point, often only produce it on the calyx of the primordial
flower, which terminates the panicle, and even there it is not
always to be found. This uncertainty renders it of much less
use than, from its apparent value, it ought to possess.

Arrhenius and Godron state that the petals furnish most
valuable characters. It unfortunately happens that they
have not received so much attention in England as it is probable that they deserve. They are sometimes very broad
so that their edges overlap; or may be so narrow as to be
quite separate from each other, and to give a star-like appearance to the flower: they are broad to the base, or wedgeshaped; rounded at the end, or lanceolate; entire, or notched
at the end; wavy at the edges, or throughout, or plain. In
colour they are most frequently white, although often pink,
or even sometimes reddish.

The colour of the filaments and anthers and styles deserves much more notice than we have given to it. It is apparently constant in tint, although very variable in intensity.

The fruits are formed of many small drupes placed close together and usually cohering. They are seated upon a receptacle, which is conical in all except the herbaceous species, and either falls with the fruit, or remains attached to the stalk after the fruit has separated from it, but amongst our species this latter condition exists in the *Idœi* alone. I have not found that the seeds afford any characters of importance. As Arrhenius places confidence in their form, it is desirable that attention should be paid to them[1].

If a bramble is found to retain the same appearance under different circumstances of soil and exposure, although many of its characters vary considerably, we may conclude that it is a true species, and form some idea of its range of variation: but when a plant, although furnished with rather marked characters is confined to a single spot, we properly doubt its specific claims, although necessity may oblige us to allow it to stand alone, from not knowing with what other plants it should be combined. No study in herbaria can supply the knowledge requisite for a determination of the

---

[1] The following very curious description of the fruit of *Rubus* is to be found in *Linnœi Amœnitates Academicœ* (viii. 170): "*Rubus*, fructu suo singularis admodum est: Receptaculum enim seminum quasi duplici epidermide obducitur, intra quam succus et semina latent, adeoque decidit hæc bacca concava instar pilei." It is probably the description of the student (Sveno Anders Hedin, who defended it under the presidency of Linnæus on May 26, 1772, at Upsala), not of the Professor. Linnæus adopted the genus from Tournefort, who described the fruit correctly in his *Institutiones Rei Herbariæ* (ed. 3. 614. t. 385).

range of variation in these plants. Unfortunately our
information on this subject is rarely sufficient to give confidence to our determinations. The recorded geographical
distribution of a species is often far from telling the whole
truth : it may seem to show that a plant is confined to a single
spot, or nearly so, and thus cause just doubts concerning its
being a distinct species; whereas, in reality, it is so abundant in that place, and under such various circumstances,
that its claim to be considered as a distinct species may be
held to be well founded. For instance, *R. pyramidalis* seems,
by the geographical table, to have been found in three or
four localities, separated by long distances, and would probably have been considered as a doubtful species, had not
its extreme abundance in the valley of Llanberis attracted
especial attention to it, and shown that its limits of variation are narrow, and that it presents a clearly distinct appearance (facies), and also admits of an accurate definition.

Some botanists have ventured to state that the seeds of
Brambles do not readily germinate, that therefore we sel-
dom see a seedling, and that thickets of these plants are
almost wholly derived from the rooted ends of the stems.
Careful observation has proved to me that the exact opposite is the fact, that the seeds germinate freely, and that
seedlings are easily found in abundance by those who search
for them in the proper places.

Mr H. C. Watson informs me that Brambles are sown
by the birds in his grounds at Thames Ditton, and that
abundant seedlings appear, and have to be carefully removed; and that that is also the case in his hedges, which
he has known from the time of their being planted more
than thirty years since. During the whole of that time

seedling *Rubi* have frequently sprung up here and there in the hedgerows, although they are never allowed to fruit, and the roots are removed every winter as completely as possible.

More than forty of the supposed species have been raised from seeds in the Cambridge Botanic Garden, and the produce has not varied in form or characters from the parent plants. The seeds were sown in the autumn, and the young plants usually appeared in the succeeding May or June. The seedlings have two little oval cotyledons, and produce a small cluster of simple leaves in the course of the first summer. In the second summer short slender shoots spring from the terminal bud and the axils of the leaves in the cluster and bear ternate or sometimes a few quinate leaves. In the third summer these shoots bear small panicles; and the root throws up the strong stems of adult plants, which, in the fourth summer, bear the perfect panicles proper to the species. Although most of the stems die down to their base after they have produced panicles, that is far from being constantly the case when the stem has not succeeded in rooting at its end. It may continue to live for many years, throwing out secondary and tertiary stems, which bear panicles. But when it has rooted, only the lower part seems able to survive the succeeding winter, and the new plant formed at its end becomes detached.

On the other hand, some persons fancy that the inclination of these plants to produce fertile seeds is so strong as to result in abundant hybridity; and by that, combined with increase of the individual by offsets, they account for the many forms which are found in the genus. It is my belief that this latter view is also unfounded; and that the production of hybrids is as repugnant to Brambles as it is to most other plants. Those who adopt this view make no attempt at proof. As has been already remarked, the assumption of hybridity in difficult cases seems merely a mode of escape from, not the removal of, a difficulty. It is often nothing more than the concealment of ignorance under a bold exterior. I believe in the distinctness of species, although unable to demonstrate it. The great length of time requisite for experimental proof, the only kind which could result in a real demonstration, renders the absolute determination of this problem nearly impossible. Perhaps the most extreme instance of the attempt to explain everything by hybridity will be found in the *Reform Deutscher Brombeeren* of Otto Kuntze (Leipzig, 1867), where all the recorded German brambles are reduced to *R. fruticosus* L. (= *plicatus, affinis,* and *nitidus,* of W. and N., and *corylifolius* of Hayne), *R. candicans* Weihe (=*fruticosus* W. aud N., and *thyrsoideus* Wimm. in part), *R. sanctus* Schreber (= *discolor, villicaulis, carpinifolius,* and *Schlechtendalii* of W. and N.), *R. Idæus* L., *R. cæsius* L., *R. Radula* Weihe, *R. hybridus* (= *pygmæus, glandulosus, Koehleri, Hystrix, humifusus, rosaceus,* and a host of others), *R. saxatilis,* and *R. Chamæmorus.* In addition there are 23 supposed hybrid plants: but in many cases the supposition seems to me to be very rash, for in this country the supposed parents have not been observed growing in company.

Dr Lejeune and also M. Alexis Jordan have cultivated brambles extensively and raised them from seeds. They find that the character of the species continues constant even after repeated sowings. The Abbé Chaboisseau justly remarks that it would require a century or more to be spent in experiments by cultivation from seed to attain to any certain result. He adds: 'L'habitant des grandes villes, condamné à étudier beaucoup plus en herbier que sur le nature vivante, se fait de l'espéce une idée tout autre que l'observateur placé au milieu des champs. Il se forme de chaque espèce un type ideal plus au moins large, selon le nombre et les états des spécimens qu'il a pu voir dans les herbiers." *L'Étude specifique du genus Rubus.*

In the attempt that is made to point out the geographical distribution of the species I have been obliged to trust chiefly to my own collection for information; for in the present uncertain state of the nomenclature of brambles it is not advisable to accept the names given even by the best botanists. The tables show the presence of the several species in certain parts of the country; but do not, and cannot, point out their abundance or rarity in any place. This is an unfortunate circumstance, for, as has been already indicated, much depends upon it. In illustration: *R. discolor* and *R. Radula* are equally marked as natives of Prov. III. *Ouse* and county of Cambridge. The former is exceedingly abundant; the latter has only been found in one place. It seems probable that one or more of the species constitutes the prevalent bramble, the Blackberry, of each district. *R. discolor*, which is very common in many parts of the kingdom—so abundant as to attract notice almost exclusively to itself—is superseded by another kind in some places, where it may be and probably is present, but escapes general notice. *R. diversifolius* is so abundant in the valley of the Severn at and for a long distance above Shrewsbury, and *R. pyramidalis* and *R. incurvatus* at Llanberis, as to be noticed by any observant person; but *R. discolor* is not seen except by the botanist who is familiar with brambles.

The eighteen *Provinces* into which Mr H. C. Watson divided Great Britain, and which are used in his *Cybele*

Britannica, are here adopted with the same numerical arrangement as is employed by him. His 112 Counties and Vice-counties (*Cyb. Brit.* iii. 526—528) are also used. It has likewise seemed desirable to give such imperfect information as has been obtained relative to the distribution of these plants in Ireland; 1 have therefore divided that country in a similar manner. This division of Ireland was first proposed in a communication read to the "Dublin University Zoological and Botanical Association," and published by that body in the original (Dublin) *Natural History Review* (vi. Pt. 2. 533), and in the *Proceedings* of the Association (i. 246); but as those works may not everywhere be easily accessible, a list of the *Provinces* and *Counties* or *Vice-counties* of Ireland is subjoined. The numbering is continuous from Mr Watson's similar divisions (*Cybele Britannica,* iii. 526).

PROVINCES.

XIX. South Atlantic.	XXV. Upper Shannon.
XX. Blackwater.	XXVI. North Atlantic.
XXI. Barrow.	XXVII. North Connaught.
XXII. Leinster Coast.	XXVIII. Erne.
XXIII. Liffey and Boyne.	XXIX. Donegal.
XXIV. Lower Shannon.	XXX. Ulster Coast.

COUNTIES AND VICE-COUNTIES.

XIX. SOUTH ATLANTIC.	XX. BLACKWATER.
113. South Kerry.	116. North Cork.
114. North Kerry.	117. Waterford.
115. South Cork.	118. South Tipperary.

XXI. Barrow.
119. Kilkenny.
120. Carlow.
121. Queens.

XXII. Leinster Coast.
122. Wexford.
123. Wicklow.

XXIII. Liffey & Boyne.
124. Kildare.
125. Dublin.
126. Meath.
127. Louth.

XXIV. Lower Shannon.
128. Limerick.
129. Clare.
130. East Galway.

XXV. Upper Shannon.
131. North Tipperary.
132. Kings.
133. West Meath.
134. Longford.

XXVI. North Atlantic.
135. West Galway.
136. West Mayo.

XXVII. North Connaught.
137. East Mayo.
138. Sligo.
139. Leitrim.
140. Roscommon.

XXVIII. Erne.
141. Fermanagh.
142. Cavan.
143. Monaghan.
144. Tyrone.
145. Armagh.

XXIX. Donegal.
146. Donegal.

XXX. Ulster Coast.
147. Down.
148. Antrim.
149. Derry.

As a few of the larger counties are divided into two Vice-counties the lines used for that purpose must be described.

Kerry is divided into North and South by a line which follows the course of the river Flesk from the place where it enters the county to its mouth in the Lower Lake of Killarney, then skirts the northern shore of that lake as far as the river Laune, which it descends to the sea. *Cork*

is separated into North and South by a line descending
the river Sullane from its entrance into the county to its
junction with the river Lee, and then following the course
of that river to the sea. *Tipperary* is conveniently divided
into North and South by the Great Southern and Western
Railway. In *Galway* the division into East and West
is well defined by Lough Corrib and the river which flows
from it. In *Mayo* a boundary between the East and West
is also tolerably well marked by Lough Mask and the course
of the river Ayle as far as a small lake above Ballyhean
church; from thence it is imaginary for a short distance
until it reaches the road from Tuam to Castlebar close to a
hamlet called Tully; then it follows that road as far as
Castlebar, and from thence descends the course of the water
through Lough Cullin and by the river Moy to the sea
near Ballina.

This division into Provinces has been adopted by Messrs.
Moore and More in their *Cybele Hibernica*, but they have
not thought it desirable as yet to attempt determining the
distribution of the plants under the Counties separately.

One inconvenience of the tabular form has already been
mentioned, another is that it does not give a very satisfactory idea of the extent to which the country has been
examined. For instance, Provinces vi. *South Wales*, and
vii. *North Wales*, seem to be tolerably known; but in fact
only very small parts of them are in that condition. Few
or no *Rubi* are recorded from the counties of Glamorgan,
Brecon and Caermarthen, in S. Wales; or from Denbigh,
Flint and Anglesea, in N. Wales. Also it is only some
small parts of the other counties that have been examined;
viz. spots where a botanist interested in brambles has
been able to reside for a considerable time.

The Fruticose Rubi do not ascend to a great elevation above the level of the sea. Mr Watson (*Compend. of Cyb. Britan.* 19) considers their upper limit to be in his *Super-agrarian zone*, which is characterized by the presence of *Quercus*, *Fraxinus*, *Lonicera* and *Crataegus*, and by the presence of *Pteris* without *Rhamnus*.

In the vale of Llanberis in North Wales 600 feet is about the height at which they appear to cease. Below that elevation they are immediately plentiful; above that I only noticed one bush (*R. discolor*), which was growing under a wall at the great height of 1000 to 1100 feet; but as no others occurred, its existence there was probably the result of accident. Mr Lees (*Bot. of Worcest.* p. 142) states that "In general *Rubi* delight in hilly spots of moderate height, becoming prostrate...at above 2000 feet of elevation, but descending and luxuriating even on the sands of the sea-shore." It is therefore possible that the elevation which my observation has led me to adopt as the upper limit of their growth is too small, especially as Mr H. C. Watson gives 900 ft. as the highest point at which they are found in the West Highlands of Scotland, and Mr Baker about the same elevation in the Humber and Tyne Districts. It would be interesting to ascertain if the *R. suberectus* which Mr Lees informs me that he found near "Gors Lwm on Banwen" Mountain in Glamorganshire, at an elevation of about 2000 feet, is the true plant, or is not rather the *R. fissus*. I have never noticed *R. suberectus* on exposed spots such as that must be, but have often seen *R. fissus* on open mountain sides, although never at so great an elevation. *R. fissus* is the *R. suberectus* of many recorded stations, especially of those in Scotland.

To judge properly of a bramble from a preserved specimen we require a piece of the middle of the stem with more than one leaf; the base and tip of the stem are also desirable. Likewise a piece of the old stem with the flowering shoot attached to it; the panicle with flowers, and the fruit. We likewise want to know the direction of the stem throughout, of the leaflets, and of the calyx; also the shape of the petals and the colour of the styles: a note of these should be made when the specimen is gathered.

In quoting the works of different authors I do not hold myself responsible for the correctness of all the synonyms given by them. In some cases I have no doubt of their incorrectness, but do not possess any absolute proof of it.

The localities for each species are with comparatively few exceptions founded upon specimens preserved in my own Herbarium. When such is not the case the authority is added within brackets and a (!) appended wherever I have seen a specimen. But as many of these latter have not been recently seen I must not be considered as now guaranteeing their absolute accuracy.

BOOKS QUOTED IN THIS VOLUME.

———◆———

Anders.— . Anderson, G.   In Linnean Transactions, xi. 4to. London, 1815.

A. N. H.— . Annals of Natural History, 8vo. London, 1838, &c.

Arrh.— . . Arrhenius, Monographia Ruborum Sueciæ, 8vo. Upsala, 1840.

Bab.— . . Babington, Manual of British Botany, 12mo. London, Ed. 1, 1843; Ed. 2, 1847; Ed. 3, 1851; Ed. 4, 1856; Ed. 5, 1862; Ed. 6, 1867.

Primitiæ Floræ Sarnicæ, 12mo. London, 1839.

Synopsis of the British Rubi, 8vo. London, 1846 (also in Ann. Nat. Hist. xvii., and Transactions of the Edinburgh Botanical Society, ii.).

In Annals of Nat. Hist. xix., and Ser. 2. ix. (also in Trans. Edin. Bot. Soc. iii. and iv.).

In Botanical Gazette, i. 8vo. London, 1849.

Flora of Cambridgeshire, 12mo. London, 1860.

Bell Salt.— .   Bell Salter. In Ann. of Nat. Hist. xv. and xvi. 1845.

In Phytologist, ii. 1845.

In Botanical Gazette, ii. 1850.

In Hooker and Arnott's British Flora, 12mo. London, 1850.

In Bromfield's Flora Vectensis, 8vo. London, 1856.

Blox.— . .   Bloxam. In Kirby's Flora of Leicestershire, 12mo. 1850.

Boenn.—   .   Boenninghausen, Prodromus Floræ Monasterensis Westphalorum, 8vo. Monast. 1824.

Bor.—   . .   Boreau, Flore du Centre de la France, Ed. 3, 8vo. Paris, 1857.

Borr.— . .   Borrer. In Hooker's British Flora, 8vo. London, Ed. 2, 1831 ; Ed. 3, 1835.

Bot. Gaz.— .   The Botanical Gazette, 8vo. London, 1849 —51.

Chab.— . .   L'Abbé Chaboisseau, De l'étude specifique du Genus Rubus (Congrès Sc. de France, 28 Session, tom. iii.), Bordeaux, 1863.

E. B.— . .   Smith and Sowerby, English Botany, 8vo. London, 1791—1814.

E. B. S. . .   Supplement to the English Botany, 8vo. London, 1831, &c.

Fr.— . . .   Fries, Mantissa tertia Novitiarum Fl. Suecicæ, 8vo. Upsala, 1842.

Summa Vegetabilium Scandinaviæ, 8vo. Holmiæ, 1846.

Garke— . .   Flora von Nord- und Mittel-Deutschland, 12mo. Ed. 7, Berlin, 1865.

Genev.— . .   Genevier, Essai (1) sur quelques espèces du

Genus Rubus de Maine et Loire et de la Vendée (Mém. Soc. Academ. d'Angers, viii.), 8vo. Angers, 1860.

Essai (2) sur quelques espèces du Genus Rubus de Maine, &c. (Mém. Soc. Acad. d'Ang. x.), 8vo. Angers, 1861.

Observations sur la Collection de Rubus de l'Herbier de T. Bastard (Mém. Soc. Acad. d'Ang. xiv.), 8vo. Angers, 1863.

Godr.— . Godron. In Grenier et Godron Flore de France, 8vo. Paris, 1848.

Flore de Lorraine, Ed. 2, 12mo. Paris, 1857.

Monographie des Rubus aux environs de Nancy, 8vo. Nancy, 1843.

Hall— . In Transactions of the Royal Society of Edinburgh, iii. 4to. Edinburgh, 1794.

Hort— . . In Ann. Nat. Hist. Ser. 2, vii. 8vo. London, 1851.

Johns.— . . Johnston, Botany of the Eastern borders, 8vo. London, 1853.

Lam.— . . Lamarck, Flore Français, Ed. 1, 8vo. Paris, 1778.

Lange— . . Danske Flora, 12mo. Kjobenhavn, 1856—59.

Lees— In Steele's Handbook of Field Botany, 12mo. Dublin, 1847.

In Phytologist, iii. 8vo. London, 1848.

Botany of Malvern, Ed. 2, 12mo. London, 1852.

Botany of Worcestershire, 8vo. Worcester, 1867.

Leight.— . Leighton, Flora of Shropshire, 8vo. Shrewsbury, 1841.

In Phytologist, iii. 8vo. London, 1848.

Lej.— . . . Lejeune, Review de la Flore de Spa, 8vo. Liege, 1824.

Lindl.— . . Lindley, Synopsis of the British Flora, 12mo. London, Ed. 1, 1829; Ed. 2, 1835.

Linn.— . . Linnæus, Species Plantarum, 8vo. Holmiæ, Ed. 1, 1753; Ed. 2, 1762.

Flora Sueciæ, 8vo. Ed. 2, Stockholm, 1755.

Merc.— . . Mercier, Rubi Genevienses (in Reuter, Cat. Pl. de Genève, Ed. 2, 1861), 12mo. Genève.

Metsch— . . Rubi Hennebergenses. In the Linnæa, xxviii. (xii.), 8vo. 1856.

Müll.— . . Müller, Versuch einer Monographischen darstellung der gattung Rubus, in the Pollichia, 1859.

Peterm.— . Petermann, Flora Lipsiensis, 12mo. Lipsiæ, 1838.

Phytol.— . The Phytologist, 8vo. London, 1841, &c.

Poir.— . . Poiret, Encyclopédie Méthodique. Botanique, 4to. Paris, 1783—1817.

Ray— . . Synopsis Methodica stirpium Britannicarum, 8vo. London, Ed. 1, 1690; Ed. 2, 1696; Ed. 3, 1724.

Reichenb.— . Reichenbach, Flora Germanica Excursoria, 32mo. Lipsiæ, 1830.

Rub. Germ.— Weihe et Nees' v. Esenbech, Rubi Germanici, Fol. Elberfeld, 1822—27.

Schultz— . Prodromus Floræ Stargardensis, 8vo. Berlin, 1806.

Supplementum ad Fl. Stargard. 8vo. Neobrandenberg, 1827.

Archives de la Flore de France et d'Alle-
magne, 8vo. Bitche, 1842—55.

Ser.— . . Seringe. In De Candolle, Prodromus, ii.
8vo. Paris, 1825.

Sm.— . . . Smith, Flora Britannica, 8vo. London, 1800.
English Flora, 8vo. London, 1824.
English Botany, 8vo. London.

Sonder— . Flora Hamburgensis, 12mo. Hamburg, 1851.

Syme . . . English Botany, Ed. 3, 8vo. London, 1863,
&c.

Trattin.— . Trattinnick, Rosacearum Monographia,
12mo. Vindebonæ, 1823.

Van den Bosch.—Prodromus Floræ Batavæ, 8vo. 1850.

Wallr.— . . Wallroth, Schedulæ Criticæ, 8vo. Halæ,
1822.

Webb and Flora Hertfordiensis, 12mo. London,
Coleman.— 1849.

Weihe— . . In Bluff et Fingerhuth, Compendium Floræ
Germanicæ, 12mo. Norinbergiæ, Ed. 1, 1825;
Ed. 2, 1837.
In Wimmer et Grabowski, Flora Silesiæ,
8vo. Vratislaviæ, 1829.

Wimm.— . Wimmer, " Flore von Schlesien, 8vo. Berlin,
1832."
Flore v. Schlesien, Preuss. und Osterr. An-
theils, 12mo. Breslau, 1840.

Woods— . Tourist's Flora, 8vo. London, 1850.
In Phytologist, Ser. 2, 8vo. London, 1856.

THE BRITISH RUBI.

Nat. Ord. *ROSACEÆ.*

Subord. ROSEÆ.   Tr. DRYADEÆ.

Rubus Linn.

Calyx explanatus, limbo 5-partitus, ab ovariis discretus, persistens. *Petala* quinque, calyci inserta. *Stamina* indefinita, cum petalis inserta. *Ovaria* plura, receptaculo convexo imposita, unilocularia. *Stylus* subterminalis, filiformis, brevis; stigma simplex. *Acini*[1] succosi, receptaculo protuberante exsucco impositi, monospermi, basi inter se confluentes. *Semen* inversum, prope basin styli affixum ; radicula supera.

Stems herbaceous, or more frequently rather shrubby, erect, or ascending or trailing, generally prickly, leafy, angular or nearly round, often rooting at the end, rarely more than biennial. *Leaves* alternate, stalked, digitate or impari-pinnate or ternate, rarely simple. *Stipules* adnate to the petiole or springing from it. *Flowers* terminal and

---

[1] *Acinus* est bacca mollissima, succulenta, subtransparens, constanter unilocularis, et seminibus duris, uno aut pluribus referta. Gaert. *de Fructibus*, v. 1. xcvi.

axillary, in the shrubby species forming racemes, panicles or corymbs, which spring from shoots produced by the stems of the preceding year. *Calyx* without bracteoles. *Petals* deciduous, white or reddish. *Inflorescence* centrifugal.

Sec. I. Rubi Frutescentes.

Caules suffruticosi. *Folia* subquinata. *Stipulæ* lineares ad basin petiolorum affixæ. *Flores* subpaniculati. *Acini* in baccam polyspermam compositam congesti. *Receptaculum* conicum.

Subsec. I. Idæi. *Caules* steriles suberecti biennes. *Receptaculum* a fructu discretum. *Folia* sæpissime pinnata.

1. R. Idæus *Linn.*
2? R. Leesii *Bab.*

Subsec. II. Fruticosi. *Caules* biennes vel perennantes. *Receptaculum* ad fructum adherens et cum eo desidens. *Folia* digitata, pedata, vel rarissime subpinnata.

i. *Suberecti.* Caules sæpissime suberecti, glabri vel sparsim pilosi vel pruinosi, nec setosi neque tomentosi. Aculei æquales.—Sepala intus albo-tomentosa, extus pilosa margine albo-tomentosa.

3. R. suberectus *Anders.*
4. R. fissus *Lindl.*
5. R. plicatus *W. & N.*
6. R. affinis *W. & N.*

ii. *Ramnifolii.* Caules plus minusve arcuati, sparsim pilosi, nec pruinosi nec setosi neque tomentosi, radicantes. Aculei in caulis angulis sæpissime congesti, subæquales.—Sepala extus tomentosa.

7. R. Lindleianus *Lees.*
8. R. rhamnifolius *W. & N.*
9. R. incurvatus *Bab.*
10. R. imbricatus *Hort.*
11. R. latifolius *Bab.*

iii. *Villicaules.* Caules plus minusve arcuati, pilosi vel calvati, sæpe tomentosi, glandulis subsessilibus; vel raro setosi aciculatique. Aculei in caulis angulis congesti, subæquales; vel paucis minoribus sparcis. Foliola infima petiolata intermediis dissita (*R. Grabowskii* excepto).

a. *Discolores.* Caulis aculei æquales validi, pubescentia arcte adpressa. Folia subtus cano-tomentosa.

12. R. discolor *W. & N.*
13. R. thyrsoideus *Wimm.*

b. *Sylvatici.* Caulis aculei æquales mediocres, pili patentes sæpe densi. Folia subtus virides vel raro cano-tomentosa.

14. R. leucostachys *Sm.*
15. R. Grabowskii *Weihe.*
16. R. Colemanni *Blox.*
17. R. Salteri *Bab.*
18. R. carpinifolius *W. & N.*
19. R. villicaulis *W. & N.*
20. R. macrophyllus *Weihe.*

c. *Spectabiles.* Caulis aculei subæquales, setæ et aciculi breves perpauci, pili sæpe densissimi.

21. R. mucronulatus *Bor.*
22. R. Sprengelii *Weihe.*

4

d. *Radulæ.* Caules punctis elevatis rigidis, ubi setæ aciculique breves subæquales sederunt, asperi efficiuntur; aculei subæquales.

23. R. Bloxamii *Lees.*
24. R. Hystrix *Weihe.*
25. R. rosaceus *Weihe.*
26. R. pygmæus *Weihe.*
27. R. scaber *Weihe.*
28. R. rudis *Weihe.*
29. R. Radula *Weihe.*

iv. *Glandulosi.* Caules arcuato-prostrati vel prostrati, radiantes, hirti. Aculei copiosi, valde inæquales, sparsi, in aciculos setasque copiosos graduatim adeuntes.

a. *Koehleriani.* Folia quinata vel raro ternata. Aculei aciculi setæque ad basin incrassati.

30. R. Koehleri *Weihe.*
31. R. fusco-ater *Weihe.*
32. R. diversifolius *Lindl.*
33. R. Lejeunii *Weihe.*

b. *Bellardiani.* Folia ternata vel raro quinato-pedata; foliola infima intermediis dissita, pedicellata. Aculei in caulium aciculatorum setosorum valde hirtorum angulis sæpissime congesti.

34. R. pyramidalis *Bab.*
35. R. Guntheri *Weihe.*
36. R. humifusus *Weihe.*
37. R. foliosus *Weihe.*
38. R. glandulosus *Bell.*

v. *Caesii.* Caules sæpissime arcuato-prostrati, teretes vel subangulati, pruinosi. Aculei inæquales. Aciculi setæ pilique pauci vel nulli.

39. R. Balfourianus *Blox.*
40. R. corylifolius *Sm.*
41. R. althæifolius *Hort.*
42. R. tuberculatus *Bab.*
43. R. cæsius *Linn.*

SEC. II. R. HERBACEI.

Caules herbacei. Folia ternata vel simplicia. Stipulæ ovatæ cum petiolum caulem amplectens. Flores umbellati vel subsolitarii. Receptaculum planum.

Subsec. I. SAXATILES. Caules flagelliformes. Flores umbellati vel subsolitarii. Acini magni, pauci, discreti.

44. R. saxatilis *Linn.*

Subsec. II. ARCTICI. Caules steriles nulli. Rhizomata subterranea longa. Flores terminales, subsolitarii. Acini in baccam compositam congesti.

45. R. Chamæmorus *Linn.*
[R. arcticus *Linn.*]

RUBUS Linn.

Sec. I. Rubi Frutescentes.

Caules suffruticosi. Folia subquinata. Stipulæ lineares ad basin petiolorum affixæ. Flores subpaniculati. Acini in baccam polyspermam compositam congesti. Receptaculum conicum.

Subsection I. Rubi Idæi.

Caules steriles suberecti, biennes. Receptaculum a fructu discretum. Folia sæpissime pinnata.

The fruit being so constructed that the receptacle is not deciduous with the acini separates the *Idæi* from the *Fruticosi* in a marked manner. These may be considered as the only truly erect brambles included in our flora; for their stems have never been even suspected of arching so as to reach the ground at the end, and rooting there. But although these completely erect stems of *R. Idæus* and its allies are very characteristic of the plants contained in this group, nevertheless, there are some other species which have suberect stems; especially those of the group denominated *Suberecti.* But the peculiar characteristic of the *Idæi* consists in their fruits separating from the receptacle, whereas in all the *Fruticosi* the receptacle falls with the acini adhering to it.

4—3

1. R. Idæus Linn.

R. caule erecto tereti pruinoso, aculeis setaceis rectis, foliis quinato-pinnatis ternatisve subtus niveo-tomentosis, foliolo terminali longè pedicellato *lateralibus dissitis, aculeis ramorum floriferorum* et pedunculorum *è basi dilatata compressa deflexis,* floribus racemosis.

R. Idæus Linn.! Sp. Pl. ed. 1. 492 (1753); Fl. Suec. ed. 2. 172. Sm.! Fl. Br. ii. 541; Eng. Fl. ii. 407. Eng. Bot. t. 2442. Trattin. Ros. iii. 6. Rubi Germ. 107. t. 47. Lindl.! Syn. ed. 1. 95; ed. 2. 92. Borr.! in Hook. Br. Fl. ed. 2. 243; ed. 3. 245. Arrh. Mon. 11. Leight.! Fl. Shrop. 222. Johnst.! E. Bord. 60. Fries Summa, 164. Bab.! Man. ed. 6, 105; Syn. 6. Godr. Mon. 40; Fl. Lorr. ed. 2. i. 245; Fl. de Fr. i. 551. Lees! in Steele, 60; Bot. Malv. 58. Blox.! in Kirby, 48. Sond. Fl. Hamb. 271. Bor. Fl. centre Fr. ed. 3. 187. Metsch in Linnæa, xxviii. 104. Garke Fl. Deutschl. ed. 7. 127. Syme's Eng. Bot. iii. 160. t. 442. Wirtg.! Herb. Rub. 115 and 47 (sp.). Billot! Fl. Gall. et Germ. exsic. 1658 (sp.).

R. frambœsianus Lam. Fl. Fr. ed. 1. iii. 135 (1778).

R. Idæus spinosus fructu rubro Raii Syn. ed. 1. 228; ed. 3. 467; Hist. 1640.

Rhizome creeping. *Stem* erect, nodding at the top, pruinose, terete, downy, 2-6 feet high. *Prickles* usually many, setaceous, declining, purple, sometimes few, from an enlarged base of the same colour. *Leaves* quinate-pinnate or ternate. *Leaflets* plicate, downy above, snowy white and felted beneath, lobate-serrate; lower pair subsessile, broadly ovate, acuminate, sometimes lobed on the outer edge; upper pair distant from the lower, sessile, ovate, acute, rather un-

equal below; terminal stalked, ovate, subcordate at the base,
acuminate; stipules slender; petioles which are channelled
above and under side of midribs with a few small slightly
hooked prickles. Rarely the leaves have three pairs and
a terminal leaflet.

Flowering shoots surrounded at the base by fuscous
scales, short, flowering throughout their length. *Prickles*
small, deflexed, from a compressed dilated base, coloured
like those of the stem, sometimes very few. *Leaves* ternate,
rarely pinnate. *Leaflets* like those of the stem. *Peduncles*
from the lower axils one- or few-flowered. *Panicle* many-
flowered, corymbose; flowers all pendulous. *Sepals* ovate,
acuminate, with a slender reflexed point, greenish white,
felted, with a white edge, often prickly, spreading. *Petals*
narrow, erect, white. *Fruit* crimson or amber-coloured.

There is a variety of this plant having amber-coloured
fruit, pale prickles on the stem, and the leaflets rather obovate. It is the White Raspberry of gardens, but is not often
found wild. The *R. Idæus β. asperrimus* (Lees ! in Steele's
Handb. 60) is a very prickly state of this variety. His
specimen is trifoliate.

A form having septenate leaves on the stem and pinnate
leaves on the flowering shoots is mentioned by authors, but
I have not seen a specimen of it.

The British ternate-leafed plant (*β. trifoliatus* Bell Salt.!
in *Ann. Nat. Hist.* xvi. 365 ; and Bromf. *Fl. Vect.* 154) is
very strong and luxuriant. Its leaves are larger than those
of any other form of the species ; the terminal leaflet is
long-stalked, deeply cordate at the base, and often has three
deep acuminate lobes at the end. I possess specimens of it
from the Isle of Wight and the Lake Country of the north
of England. I have not seen the *γ. microphyllus* of Lees
(l. c.), which he describes as possessing very small trifoliate
leaves, but gives no further information concerning it. It

is probably the *β. microphyllus* of Wallroth (*Sched.* 226), which has "foliis constanter ternatis, duplo minoribus; caule 1- 2-pedali, recto, a basi inde ramoso." A trifoliate specimen, gathered at Shrawley in Worcestershire in the year 1836, and supposed by Mr Lees to be "probably *var. trifoliatus* of Bell Salter and Babington," is not that plant. Its leaflets, although only in threes, closely resemble those of the typical *R. Idæus* both in size and shape. Such is also the case with his *var. asperrimus* mentioned above.

Habitat.—Damp edges of woods, and heaths. June.

Area.— 1 2 3 4 5 6 7 8 9 10 11 12 13 14 15 16 . 18 19 20 21 22 23 24 25 26 27 28 29 30.

Localities.—It is so generally distributed that no special localities are requisite. I have not seen specimens from (xvii.) the *North Highlands*, nor (xviii.) the *North Isles*, but Mr Watson states that it was found by the late Dr Neil in Orkney. It is recorded in the *Cybele Hibernica* as occurring in Provinces 23, 25, 26, 27 and 29.

2? **R. Leesii.** Bab.

R. caule erecto tereti, aculeis setaceis rectis, *foliis ternatis*, foliolis omnibus rotundato-ovatis subsessilibus imbricatis, *aculeis ramorum floriferorum* pedunculorumque paucis setaceis *basi bulbosis*, floribus racemosis.

R. Leesii Bab.! in Steele's Handb. 60 (1847); Man. ed. 3. 92 (1851); ed. 6. 105; in Ann. Nat. Hist. Ser. 2. ix. 123. Lees in Eng. Bot. Suppl. t. 2981. Syme's Eng. Bot. iii. 161. t. 443.

R. Idæus γ. *Leesii* Bab.! Syn. 6 (1846). Bell Salt. in Hook. and Arn. Br. Fl. ed. 7. 123.

R. Idæus c. *fragrariæ-similis* Lees in Lond. Cat. Br. Pl. (name only).

Rhizome widely creeping. Stem erect, nodding at the top, terete, downy, with short adpressed hairs and many subsessile glands, 2—3 feet high. *Prickles* many, slender, setaceous, declining, with bulbous or oblong bases, pale purple. *Leaves* ternate. *Leaflets* similar, roundly ovate, dark green and rugose above, snowy white and felted beneath, coarsely crenate-serrate-apiculate; lateral sessile, overlapping the very shortly stalked terminal leaflet. *Petioles* furrowed and having a few small declining prickles beneath; the midribs of the leaflets either unarmed or having a very few minute prickles beneath. *Stipules* very slender.

Flowering shoots from fuscous scales, short, flowering in their upper half, clothed with short deflexed hairs. *Prickles* slender, setaceous, declining, from bulbous bases. *Leaves* mostly simple, cordate, slightly 3-lobed, very coarsely crenate-serrate, green above, greenish white beneath: sometimes there is a small ternate leaf at the base of the shoot

having small roundly obovate blunt leaflets, all very shortly
stalked, but the stalk of the terminal leaflet rather the
longest. *Flowers* in a lax simple raceme of which one or
two of the lowest peduncles are axillary. *Peduncles* with
small and very slightly deflexed prickles. *Sepals* oblong,
often more than five in number, in which case they are
linear, cuspidate, greenish white and felted on both sides;
the point slender, short, slightly reflexed, glabrous. *Petals*
spathulate, acute, white. *Stamens* and *styles*, white. *Fruit*
rarely produced, of few crimson drupels with the taste of
Raspberry, and doubtfully perfect. One or two drupels
gathered in 1865, in the Cambridge Botanic Garden, seemed
to contain seeds.

Plants which creep extensively underground often do not
produce much fruit; but, bearing in mind that the close
ally of this plant (*R. Idæus*) fruits abundantly, the fact that
R. Leesii rarely attempts the formation of fruit, and that
even when its drupes are apparently well ripened they seem
to be usually devoid of any perfected seed, we are led to
suspect its distinctness as a species. Should it really be a
state of *R. Idæus* it must be considered as exceedingly
curious. All the trifoliate forms of *R. Idæus*, with perhaps
one exception, differ remarkably from *R. Leesii* by having a
very long stalk to their terminal leaflet; also, if placed side
by side with *R. Leesii*, the leaflets are seen to have very
little similarity, however difficult it may be to convey an
idea of the difference by description. The exception referred
to is the *R. Idæus* c. *anomalus* of Arrhenius (*Mon.* 14),
which is stated to have usually "folia plerumque simplicia
cordato-rotundata vel reniformia;...folia ternata imis duobus
breviter petiolatis, ...extimo mediocriter petiolato subrotundo." If we could suppose that these words are intended
to describe the leaves of the flowering shoot alone, there
would be little doubt of the identity of the *R. Leesii* with

the *var. anomalus;* but an expression of Arrhenius shows
that he had the sterile stem before him when writing the
account of his plant, for he remarks "si ex frutice sterili
judicares, hunc ad *R. Idæum* vix traheres," and we are
therefore probably right in supposing that the leaves described in the above quotation were taken from the stem
not from the flowering shoot, although those of the latter
may have been similar. He also says "Omnia basin versus
angustata," which does not apply to the leaves of our plant.
I have not observed the least approach to the simple condition of leaves on stems of *R. Leesii*, either in a wild or
cultivated state. The lateral leaflets of *R. Leesii* accord
tolerably with the description of Arrhenius. They are almost sessile but scarcely narrowed at their base: the terminal leaflet has a stalk which I have never seen to exceed
one-sixth of an inch in length, and it is usually shorter.
Nevertheless Hr. de Brugh in the *Nederlandsch Kruid-
kundig Archief* (iv. 460) decides that our *R. Leesii* is identical with the *R. Idæus c. anomalus.* Unfortunately I have
not access to Swedish specimens of the *var. anomalus.*

The figure in *Eng. Bot. Suppl.* (t. 2981), and in Syme's
Eng. Bot. (t. 443), erroneously represents the terminal
leaflet as having a considerable stalk. This is a mistake
made by the engraver of the plate which was unfortunately
not seen in time for its correction.

Habitat.—Woods and thickets. June.

Area.—1 . . . . 12.

Localities.—i. In a wood near Ilford Bridges, three miles
from Linton, *N. Devon*, where it was discovered by *Mr E.
Lees*, in 1843. On a shingly bank near Bonniton, not far
from Dunster, *W. Som.*, where it was detected by the late
Rev. W. H. Coleman, in 1849.—xii. By the side of a stream
that flows into Windermere between Troutbeck and Bowness,
Westmoreland. Mr E. Lees, in E. B. S. fol. 2981.

Subsection 2. Rubi Fruticosi.

Caules biennes vel perennantes. Receptaculum ad fructum adherens et cum eo desidens. Folia digitata, pedata, vel rarissime subpinnata.

The following subdivisions are difficult to define, and not very clearly separated in nature; but the characters given will, it is believed, usually be found to group the plants in a satisfactory manner.

Group I. Suberecti.

Caules sæpissime suberecti, glabri vel sparsim pilosi nec setosi nec pruinosi neque tomentosi. Aculei æquales.—Sepala intus albo-tomentosa, extus pilosa margine albo-tomentosa.

The stems of these plants are very characteristic of the group. They are nearly as erect as those of the *Idæi* and nod more or less at the extremity: but sometimes they form a considerable angle (say 90°) with the horizon. In no case have I seen or been informed of their arching to the ground. One or two abnormal forms, found under deep shade, which seem to be states of the plants belonging to this group, have quite prostrate stems.

The panicle is usually nearly simple and racemose or slightly corymbose; flowers large; sepals often quite green and glabrous externally, with the exception of an edging of white felt similar to that which clothes their inner side.

3. R. suberectus Anders.

R. caule erecto obtusangulo, *aculeis* raris rectis *brevibus exiguis e basi dilatata* compressa conicis ad angulos caulis congestis, foliis 3-5-7-natis, *foliolis flexibilibus planis*, foliolo terminali cordato-acuminato *infimis* subsessilibus *ramorum floriferorum basi attenuatis*, floribus racemosis vel subpaniculatis rachi et pedunculis pilosis, sepalis a fructu (atro-sanguineo) reflexis.

R. suberectus Anders.! in Linn. Trans. xi. 218. t. 16. (1816). Sm.! in Eng. Bot. t. 2572; Eng. Fl. ii. 409. Lindl.! Syn. ed. 1. 91; ed. 2. 92. Arrh.! Mon. 19. Leight.! Fl. Shrop. 223; in Phytol. iii. 72. Blox.! in Kirby's Fl. Leicestr. 48. Lees in Steele's Handb. 59. Bab.! Man. ed. 6. 106. Fries! Summa, 164; Herb. Norm. vi. 44 (sp.). Bor. Fl. Centre, ed. 3. 204. Metsch in Linnæa, xxviii. 130. Godr. Fl. Lorr. ed. 2. i. 244. Lange! Danske Flora, 341 (excl. var. β). Billot.! Fl. Gall. et Germ. exsic. No. 1178 (sp.).

R. umbrosus Lees! in Steele, 60. (1847).

R. fastigiatus Rub. Germ. 16. t. 23. Blox. in Kirby, 48? Wirtg.! Herb. Rub. 31 (sp.).

R. plicatus Leight.! Fl. Shrop. 223. Nees! in Leight. Fl. Shrop. 224.

R. pseudo-Idæus Müll.! Mon. 2.

R. nitidus Lindl.! Syn. ed. 1. 92.

R. affinis Lindl.! Syn. ed. 2. 92.

Stem thick, of a soft spongy consistence, green (or reddish in exposed situations), terete near the base, angular or even furrowed towards the top, glabrous, bearing a few subsessile glands chiefly on its lower part, erect, 3-6 feet high, ulti-

mately nodding slightly at the top. *Prickles* few (except
near the base), small, slender, usually scarcely exceeding in
length the long diameter of their large dilated base (except
at the extremity of the stem), confined to the angles of the
stem, slightly declining. *Leaves* usually quinate, sometimes
pinnate-septenate, rarely ternate. *Leaflets* large, flat when
full-grown, thin, unequally serrate, green on both sides,
slightly adpressed-pilose and shining above, paler and hairy
on the ribs beneath, acuminate; basal leaflets of the ternate
leaves lanceolate, intermediate ovate, terminal cordate-
ovate; basal leaflets of the quinate leaves lanceolate, intermediate ovate-lanceolate, terminal cordate-ovate; basal and
intermediate leaflets of the septenate leaves like those of the
quinate leaves, but in place of one terminal leaflet there are
three leaflets of which the lateral are sessile and unequally
ovate, and the middle leaflet is shortly stalked and ovate-
lanceolate ; except as above-mentioned the leaflets are all
stalked, the basal very shortly, intermediate rather shortly,
terminal long-stalked; furrowed petioles and midribs beneath
with large-based hooked prickles. *Stipules* slender.

Flowering shoot from dark brown scales, suberect, often
scarcely more than a leafy raceme, with short deflexed (but
often very few) prickles and sessile glands. *Leaves* ternate;
leaflets all ovate, rounded or narrowed not cordate at the
base; lateral subsessile, terminal stalked; uppermost leaves
simple, ovate-lanceolate. A *panicle* or raceme ; lower
flowers axillary, usually long-stalked. *Sepals* reflexed, unarmed, ovate, leaf-pointed, externally dark green and glabrous but edged with white felt. *Petals* large, obovate,
narrowed below, entire, white, much exceeding the sepals.
Filaments and *anthers* " rather fuscous." *Styles* " greenish,"
falling short of the stamens. *Primordial fruit-stalk* shorter
than the sepals. *Fruit* dark red, ultimately of a deep red
(port-wine) colour, sourish.

R. suberectus β. trifoliatus of Bell Salter is often a very large plant having enormous leaflets, but differs in no essential point from the ordinary state of the species.

The typical form of this plant cannot be confounded with any of our other species. It has the habit of *R. Idæus;* its leaves are often septenate by the separation of two leaflets from the base of the middle leaflet, they are thin, flexible and slightly pilose or quite glabrous; the petioles and rachis bear a few short hooked prickles. The inflorescence is small, of a few solitary axillary flowers, and a small open terminal raceme. The floral leaves have all their leaflets narrowed to the base, not cordate.

This is certainly the plant of Anderson, although, very probably, some specimens of *R. plicatus* and *R. fissus* were included in his idea of the species. He found it "in the wood behind the Devil's Bridge, Cardiganshire;" a densely-shaded spot where *R. suberectus* is likely to occur, but where the presence of either *R. plicatus* or *R. fissus* is very improbable.

It may reasonably be doubted if *R. suberectus* is the *R. nessensis* of Hall (*Trans. R. Soc. Edin.* iii. 20). The "full" description spoken of by Anderson, certainly is far from what we now consider such; it is as follows:—"Rubus (Nessensis) foliis quinato-digitatis, ternatis, septenisque nudis, caule subinermi, petiolis canaliculatis; stolonibus erectis biennalibus." He also tells us that the fruit which is of the "colour of the red mulberry, has a peculiar taste." That is all. He found the plant "in different places on the banks and among the woods of Loch Ness, where it could not come from the same root." I incline to the opinion that the typical plant was what is now called *R. fissus*, notwithstanding the applicability of parts of the above character to *R. suberectus*. The former species is apparently common in the Highlands of Perth and Inverness; the latter is scarce there. I do not therefore quote *R. nessensis* as a synonym of either of these plants.

Fries makes the following important remark concerning his plant. "Hic per regiones montanas sylvaticas Gothiæ totius vulgatissimus, ubi omnes sequentes fruticosi [*R. fruticosus* = *R. plicatus, R. affinis, R. thyrsoideus*] desunt, novum offert exemplum ridiculæ hodiernæ hybriditatum venationis" (*Mant.* iii. 40). The total absence of *R. plicatus* from a province in which *R. suberectus* abounds, strongly tends to prove that Dr Walker-Arnott (*Brit. Fl.* ed. 8, 123) is incorrect in combining them, without acknowledging their distinction even as varieties. The fact that *R. plicatus* and *R. fissus* are often called *R. suberectus* by Scottish collectors, will possibly explain this proceeding of that eminent botanist.

Godron, Sonder, and Boreau, quote the *R. fastigiatus* (W. and N.) as identical with this species. They are probably correct. It is represented by the *R. umbrosus* (Lees), my specimen of which accords well with the figure in the *Rubi Germanici* (t. 2); but Mr Lees does not state, and would seem rather to deny, that the barren stems are very long (15 to 20 feet) and arching, as they are said sometimes to be by the authors of the German work. They (Weihe and Nees) remark that this arch is not a constant character of their plant: "surculus.........qui primo vere germinans, primum recte ascendit, tum per tempus prolixior, pedetentim in arcum curvatum ad terram inclinat; itaque in eodem dumeto, e libero solo surgente, surculos invenies alios fere erectos, alios ad dimidiam longitudinem dependentes, alios denique qui terræ jam redditi, novas radices propellant."

On the other hand Arrhenius (*Fries Summa*) quotes the *R. fastigiatus* (W. and N.) as an undoubted form of his *R. fruticosus* (which is our *R. plicatus*), and it seems probable that he had in view a plant closely allied to *R. fastigiatus*

(Bab.), and therefore different from that of M. Boreau, but identical with the specimen in Billot's collection (No. 1177).

If these views are correct, the *Suberecti*, although nearly constantly suberect, may, under peculiar circumstances become arched, and thus, as has been already remarked, the only really suberect European species are the *Idæi*.

Anderson's plate (*Linn. Trans.* l.c.) exceedingly **well** represents the barren stem of our plant, as does the plate in *English Botany* the flowering shoot: together they constitute a good illustration of *R. suberectus*. There is a specimen named by Anderson in Edw. Forster's Herb., now in the British Museum, from a place in Scotland named Stonybyers. It is marked by Smith as the true plant, and is our *R. suberectus*.

The specimen of *R. nitidus* (Lindl.) from the Hort. Soc. Garden, is *R. suberectus* (Anders.); in his second edition he says that it is *R. affinis* of that work.

The late Mr G. Don of Forfar, found this plant before 1813, and gave it the manuscript name of *R. intermedius!* He said of it (Headrick's *Forfarshire*, Appendix 25), "a new species. It grows near the waterfall called the Reeky Linn, on the water of Isla."

Habitat.—Boggy woods and thickets. June.

Area.—1 2 3 . 5 . 7 8 . 10 . 12 13 14 15 16 . . 19 . . . . . . 26 . . . 30.

Localities.—i. Between Barnstaple and Combe Martin, *N. Dev.;* Plym Valley (Briggs!); Exeter, *S. Dev.* (Linn. Brit. Herb.); Dunster, *S. Som.*—ii. America and Apse Castle wood, *Isle of Wight;* Ashdown Forest, *E. Suss.* (Borr.!).—iii. Easney Park wood, *Herts* (Fl. Herts.); Newbury, *Berks* (Linn. Brit. Herb.); Esher, *Surr.* (Borr.!).— v. *Worcester, Warwick* and *Hereford* (Blox.); Almond Park, *Salop.*—vii. Devil's Bridge, *Card.* (Anderson); Wood near Rhaiadyr Mawddoch, and Dolgelly, *Merion.*—viii. Charnwood

Leic. (Blox.).—x. Richmond, *N. W. York.* (Linn. Brit. Herb.).—xii. *Westmoreland* and *Cumberland* (Hort).

xiii. Gouroch, *Renf.;* Jardine Hall, *Dumf.*—xiv. Grant's House, *Berw.* (Edin. Herb.).—xv. By the river Isla, *Forf.* (G. Don!); Inverarnan, Loch Lomond; Banks of Loch Tay; Callander (Linn. Brit. Herb.), *W. Perth;* Ben Lomond, *Stirl.* (Balfour).—xvi. Dunoon, *Main Arg.* (Hooker!).

xix. *Cork* (Mackay).—xxvi. Headford, *E. Galway* (Mack.).—xxx. Deer-park, Newtown Limavady, *Derry* (D. Moore).

4. R. fissus. Lindl.

R. caule suberecto vel subarcuato, obtusangulo, *aculeis crebris* tenuibus rectis vel deflexis *è basi oblonga paululum dilatata* conicis *sparsis*, foliis 5-7-natis, *foliolis plicatis*, foliolo terminali cordato-ovato infimis sessilibus ramorum floriferorum basi sæpe plus minusve gibbosis, panicula simplici racemoso-corymbosa pilosa, *sepalis fructum* (atro-sanguineum) sæpe *laxe amplectentibus*.

R. fissus Lindl.! Syn. ed. 2. 92 (1835). Leight.! Fl. Shrop. 225; in Phytol. iii. 72; Shropshire Rubi, 2. (sp.) Bab. Man. ed. 3. 93; ed. 6. 106; A. N. II. ser. 2. ix. 124.

R. plicatus Leight.! Shrop. Rubi, 3. (sp.).

R. suberectus β fissus Lange, Danske Flora, 342.

Creeping. *Stem* hard, scarcely angular at the base, bluntly angular towards the end, considerably inclined, but not arching to the ground, 1½-2 feet long, hairy, rather glaucous, with many subsessile glands. *Prickles* many, slender, usually much longer than the long diameter of their small dilated base, scattered (that is, not confined to the angles of the stem).

Leaves quinate or pinnate-septenate. *Leaflets* rather coriaceous, plicate, unequally serrate, green on both sides, pilose and dull above, paler with rather crisped shining hairs beneath (sometimes so covered with these hairs as to seem felted); basal oblong, acute, very nearly or quite sessile; intermediate ovate; terminal cordate-ovate, cuspidate; septenate leaves similar, but in place of the terminal leaflet there are three leaflets of which the lateral are oval, acute, and sessile, the terminal ovate or obovate, subcordate below.

and stalked; furrowed petioles and midribs beneath with many hooked prickles; stipules linear-lanceolate.

Flowering shoot from dark brown scales, rather hairy, with a few subsessile glands and a few scattered declining or deflexed prickles. *Leaves* ternate; leaflets oblong-ovate, rounded or narrowed at the base; basal sessile, sometimes rather gibbous on one side at the base; terminal stalked: uppermost leaves simple, ovate. *Panicle* simple, racemose-corymbose; lower flowers axillary, longstalked. *Sepals* at first patent, afterwards often loosely clasping the fruit, ovate-acuminate, shining, green, and often nearly glabrous externally, edged with white felt. *Petals* white, oval-spathulate. *Stamens* whitish. *Styles* cream-coloured. *Primordial fruit-stalk* rather short. *Fruit* bright red until very nearly ripe, ultimately of a port-wine colour.

I formerly separated many plants from *R. fissus,* and joined them to *R. plicatus,* which I am now quite convinced really belong to *R. fissus* and that *R. plicatus* must include only the strong plants which have stout hooked prickles upon their stems, and do not usually if ever, possess a lobed terminal leaflet or three leaflets in its place. As far as our plants are concerned that form of leaflet, combined with a suberect stem, seems to be confined to *R. suberectus* and *R. fissus.* The stems of *R. fissus* are truly suberect; they sometimes arch considerably, but never reach the ground and root there. When its prickles are few in number they are confined to the angles of the stem, but when more abundant they also grow upon its convex faces, and in that respect differ from those of the allied plants. The prickles resemble those of *R. suberectus,* but are much longer relatively to their bases. The flowering shoot has not the gibbous nor broadbased lateral leaflets of *R. plicatus;* nor are they so narrowed at the base as those of *R. affinis.* The erect-patent calyx is not constantly present with the fruit.

Adpressed and reflexed sepals may be found on the same panicle.

The late Mr Borrer was of opinion that *R. fissus* is distinct from *R. plicatus* and *R. suberectus.* Mr Lees considers *R. fissus* to be only "a more prickly and hairy variety" of *R. suberectus;* but a specimen of his *R. fissus* now before me is scarcely, if at all, different from *R. affinis,* and cannot possibly be the *R. fissus* of Lindley.

Dried specimens often very much resemble *R. pruinosus* (Arrh.), and might well pass for it. But *R. pruinosus* has " caulis sterilis ad saxa procumbens, sex ulnas et ultra longas, tandemque radicans," and therefore has no true relationship to our *R. fissus.*

I have been unable to identify *R. fissus* with any of the described *Rubi* of continental authors.

Mr J. Lange considers a specimen of *R. fissus,* sent to him by me, to be a variety of *R. suberectus* (*Danske Flora,* 342), and mentions that the same plant grows near Fredericia.

Mr W. Wilson's plant from Woolston Moss, which is noticed in *Leight. Fl. Shrop.* (224), is *R. fissus.* Formerly I confounded it with *R. plicatus,* but Leighton always regarded it as distinct from that species.

Habitat.—Wet and peaty ground, June to August.

Area.—1 . . . 5 . 7 8 9 10 11 12 13 . 15 16 . . . . .
. . . . 30.

Localities.—i. Ivy Bridge, *N. Devon* (Briggs!).—v. Almond Park, Twyford Vownog near West Felton, and Shawbury Heath, *Salop;* Wood east of Tintern, *W. Glouc.*—vii. Dolwyddelan (Borr.!) and Llanberis, *Caern.;* Near Lanfihangel, between Cerreg y Druidion and Ruthin, *Denb.* (Borr.!); Dolgelly, Bala, and Cwm Bychan, *Merion.*—viii. Charnwood Forest, *Leic.* (Blox.!).—ix. Carrington Moss near Sale (G. E. Hunt!), and Woolston Moss, *Ches.*—x. Kilsdale in Cleveland (Baker!), Thirsk (Hailstone), and near York, *N.E. York.*—xi. Near High Force in Teesdale, *Durh.* (Baker!). — xii. Brathay, *Westm.* (Borr.!); Threlkeld, *Cumb.*

xiii. Jardine Hall, *Dumf.;* Loch Lomond, *Dumb.* (Hailstone!).—xv. Clova, *Forf.*—xvi. Dalmally, *Main Argyle.*

xxx. By river Foyle, near Londonderry (D. Moore), *Derry;* Saintsfield, *Down.*

5. R. plicatus W. and N.

R. caule suberecto obtusangulo, aculeis validis declinatis vel deflexis e basi oblonga dilatata conicis in
angulos caulis congestis, foliis quinatis, *foliolis* plus
minusve *plicatis* tenuibus *subtus pilosis* nec tomentosis,
foliolo terminali cordato-acuminato infimis sæpissime
subsessilibus *ramorum floriferorum lateralibus rhombeo-
ovatis basi dilatatis*, floribus racemosis vel corymbosis,
rachi et pedunculis pilosis nec tomentosis, sepalis a
fructu (atro) reflexis.

R. plicatus Rubi Germ. 15. t. 1 (1822). Trattin. Ros. iii.
30. Borr.! in Eng. Bot. Suppl. t. 2714. Leight. in Phytol. iii. 73. Bell Salt.! in Bot. Gaz. ii. 117; in Bromf., Fl.
Vect. 155. Blox.! in Kirby 47. Lees! Malv. 57. Johnst.!
E. Bord. 60. Bab.! Man. ed. 1. 97; ed. 6. 106. Lange!
Danske Fl. 342. Müll.! Vers. 2. Syme, Eng. Bot. iii. 166.
t. 445.

R. nitidus Sm.! Eng. Fl. ii. 404. Johnst.! E. Bord. 61.
Billot! Fl. Gall. et Germ. exsic. No. 2668 (sp.).

R. fastigiatus Bab.! Syn. 8; Man. ed. 2. 97. Weihe in
Reichenb. Fl. exsic. No. 786 (sp.). Wirtg.! Rub. Rhen. No.
1 et 2 (sp.). Müll.! Vers. 2.

R. fruticosus Linn. Fl. Suec. ed. 2. 172 (in part). Wahlb.
Fl. Gothob. 54. Arrh.! Mon. 23. Fries! Summ., 164; Herb.
Norm. v. 51 (sp.). Reichenb. Fl. excurs. 100. Godr. in Fl.
Fr. i. 349; Fl. Lorr. ed. 2. i. 243; Monog. Rub. Nancy, 36.
Sond.! Hamb. 272 (excl. var. β). Van den Bosch Fl. Batav.
71. Bor. Fl. Centre, ed. 3. 204. Metsch in Linnæa, xxviii.
136. Garke Fl. Deutschl. ed. 7. 118. Wimm. Fl. v. Schles.
131. Billot. Fl. Gall. et Germ. exsic. No. 1177 (sp.).

R. vulgaris Leight.! Fl. Shrop. 321 (in part).

R. suberectus Reichenb.! Fl. exsic. No. 780 (sp.). Wirtg.
Fl. Preus. Rhein. 150.

R. suberectus β. plicatus Borr.! in Hook. ed. 2. 243; ed.
3. 246.

R. corylifolius Schultz Fl. Starg. 131; Suppl. 29.

R. appendiculatus Trattin. Ros. iii. 31. DC. Prod. ii.
561 (teste Questier).

R. fol. quinato-digitatis, &c. Linn. Fl. Suec. ed. 1. 148.
No. 409.

R. hamulosus Müll.! in Jahresb. Pollichia xvi. 76.

Stem bluntly angular except at the tip, usually inclining
but not truly erect, bearing many subsessile glands, hairless,
more woody than *R. suberectus* or even *R. fissus*. *Prickles*
unequal declining or deflexed, conical, springing from an
oblong base, often falcate near the top of the stem. Lower
leaves ternate, upper quinate; or rarely pinnate-septenate.
Leaflets rather thin, serrate, very nearly glabrous above,
shining, pilose but not felted, pale yellowish green beneath,
more or less plicate; basal nearly sessile, imbricate, broadly
lanceolate; intermediate shortly stalked, broadly lanceolate;
terminal leaflet long-stalked, cordate-ovate prolonged (on the
septenate leaves it is narrowed to the base and very shortly
stalked, and its lateral leaflets are elliptic but unequal-sided);
midribs and slightly furrowed petioles with small hooked
prickles; stipules linear.

Flowering shoot from brown rather silky scales, hairy.
Prickles strong, nearly straight, declining, or more often falcate, from a very large long compressed base. Leaves ternate; lower leaflets ovate, broad and gibbous below, sessile;
petiole furrowed; uppermost floral leaves simple, cordate-
prolonged. *Panicle* or raceme leafy below, sometimes
throughout, with crisped hairs and many subsessile glands;
nearly or quite without prickles in the typical plant, or

furnished with rather abundant strong hooked prickles; terminal flower more shortly stalked than the lateral flowers. *Sepals* greenish, but finely and thickly silky externally, becoming glabrous except at the edge which has a felted border of the same fine wool which lines the inner side, reflexed. *Petals* distant, white or pinkish, ovate-spathulate, twice as long as the calyx, entire. *Anthers and styles* pale cream-coloured. *Primordial fruit-stalk* equalling the sepals and oblong fruit which is of a claret or blood-red colour, but ultimately becomes quite black, and is slightly acid.

R. nitidus (W. and N.) is usually joined to *R. plicatus* by those botanists who do not accept it as a distinct species. Weihe, Godron, and Boreau ascribe an adpressed calyx to it. Weihe's words are very decided, he says, " calycis laciniæ... peracta anthesi refractæ, maturo autem fructu rursum patentes vel etiam incurvæ." Godron also lays much stress upon it and uses the words " son calice appliqué." My specimens of English plants which it is probable belong to *R. nitidus* do not show decidedly the presence or absence of this character, which I am inclined to think is inconstant in this species, as I believe it to be in *R. affinis.* But there is another character upon which much stress is laid by those who separate the plants : *R. nitidus* has a more divided and decidedly prickly panicle; or rather, it has a panicle, whilst its ally has only a raceme. It is very difficult, indeed I consider it impossible, to distinguish between these forms of inflorescence in *Rubi*. I have before me a specimen from Mr Bloxam (gathered in 1846 at Appleby Road, Twycross, and marked as " No. 1. *R. plicatus*"), where two flowering shoots are given, of which one is simply a raceme like that of typical *R. plicatus*, bearing a very few slender declining prickles as in that plant; and the other is a panicle in which 9 out of the 11 branches are themselves branched (each of the upper having 2 and the

6

lower 4 flowers), and the rachis bears many strong compressed hooked prickles (a few of which arms may also be found on the branches), such as ought to be found on *R. nitidus* according to the text of the *Rubi Germanici*, although the plate does not represent them. If these two forms of inflorescence really belonged to the same plant, and I have every reason to have confidence in Mr Bloxam's accuracy, then the character supposed by most continental botanists to be afforded by the inflorescence and its arms fails. Indeed I think that I can trace all the connecting links between the perfectly simple and almost unarmed raceme of the true *R. plicatus* and the very strongly armed inflorescence with decidedly hooked prickles of a plant from Bantry which greatly resembles *R. montanus* (Wirtg.), and which also bears very nearly a simple raceme; and also to the plants with panicles which bear strong hooked prickles belonging, I have little doubt, to the *R. nitidus* (W. and N.).

The *R. nitidus* of Smith, as shown by specimens gathered by Mr Borrer "from the same plant as those sent to Sir J. E. Smith," is *R. plicatus*, bearing a few strong hooked prickles on its inflorescence, but in other respects typical of that species.

The *R. plicatus* sent to Borrer by Mertens is our plant; the *R. nitidus* from the same botanist is also our *R. plicatus*.

I have carefully studied the descriptions given by continental botanists who distinguish *R. plicatus* from *R. nitidus* without discovering any constant distinctive character, and am therefore confirmed in my opinion that they ought not to be separated as species. In taking this view of them I am glad to find myself in accord with Arrhenius (*Mon.* 25).

It is deserving of remark that the specimen of *R. nitidus*, published by Weihe in *Reichenb. Fl. Germ. exsic.* (No. 783), has, in my copy, two pieces of stem and two leaves: one with stellate hairs on the stem, and a cordate-ovate-acumi-

nate terminal leaflet : the other now and probably originally
glabrous, and its leaflet obovate-acuminate and only slightly
cordate below. Both have well-marked stalks to the basal
leaflets. It is very doubtful if the former stem and leaf
really belong to *R. nitidus.*

Godron says that the stem of *R. nitidus* is "dressée
arquée seulement au sommet," but other authors describe it
as "arquée radicante," and in the *Rubi Germanici* it is
stated that "arcu petit terram 6—10 pedem longitudine."

I have received from Mr Baker as *R. suberectus* a specimen of what seems to be the true *R. nitidus.* It resembles *R. plicatus* in most respects, but has slightly obovate
acuminate terminal leaflets which are rather hairy beneath,
and a slightly divided panicle in one case and a simple
raceme in another. He gathered it at Thirsk, in N.E.
Yorkshire, in 1851. Other specimens, from the Isle of
Arran (Scotland), Appleby Road at Twycross, Leicestershire, and falls of the Mynach, in Cardiganshire, are apparently also the true *R. nitidus.*

A remarkable form which I think belongs to *R. plicatus*
is the *R. fastigiatus* of my Synopsis. It had quite lost the
suberect habit, its stems being long and procumbent. It
grew in deep shade, and was probably modified thereby.
The prickles on the stem are more compressed at the base.
The lower leaflets of its very large leaves are nearly sessile
and overlap the intermediate leaflets. The *R. fastigiatus*
(W. and N.) appears to be another prostrate form of *R.
plicatus.* Its prickles, although very large, are exactly
like those of *R. plicatus.* Its lower leaflets do not overlap
the intermediate pair. A specimen gathered near Keswick
by the Rev. F. J. A. Hort agrees very exactly with this,
as do also Nos. 1 and 2 of Wirtgen's *Herb. Rub. rhenan.*
(*R. fastigiatus* (W. and N.) forma 1 et 2), and a specimen
named *R. plicatus?* (vel *R. fastigiatus*) by Lange which was

found at Bronsted in Jutland. There is a specimen in *Herb. Borr.*, gathered by him at Tilgate in Sussex, and marked as "near *R. plicatus*, but apparently rooting," which from that sample alone I should have called *R. plicatus;* but as it probably rooted, or at the least was not suberect, it is almost certainly the state of that plant called *R. fastigiatus.* Dr Bell Salter has written "*nitidus*" on the label: but the plant has nothing to do with his *R. nitidus.* There are two specimens from Mertens in *Herb. Borr.* which are named *R. fastigiatus* (Weihe): if suberect they are typical *R. plicatus;* if rooting, the form called *R. fastigiatus* by Weihe and Nees. On one of them the silky coating of the sepals is much thicker and more persistent than is usual.

Sometimes strong thorny plants much resemble *R. affinis* but may be known from it by the following marks: *R. plicatus* has a pilose but not felted top to its panicle; the lateral leaflets of its flowering shoot are dilated or gibbous below; the sepals are only slightly hairy externally, chiefly at their base and tip, although the felted edge is present in both plants; the stem leaves are hairy on the veins, but never felted beneath; the *terminal leaflet is broadest near to its base*, not at about its middle as are those of *R. affinis*, and it is acuminate (or "prolonged," to use the excellent term adopted by Mr Woods) rather than cuspidate.

I have received French specimens of this species from M. Questier with the names *R. fastigiatus* and *R. suberectus*, and Genevier gives the former name to specimens of *R. plicatus, R. fissus*, and *R. suberectus.*

The *R. nitidus* of Johnston, which he considered to be "very well" represented by the plate of *R. nitidus* in the *Rubi Germanici*, appears to me to agree exactly with the *R. plicatus* of the German authors. He considered his

R. nitidus "very different from" his *R. plicatus* and what he called "a genuine bramble." If I only possessed such specimens as are in his Herbarium, I should very probably hold the same opinion, but my large series of specimens shows that they are forms of the same species. It is hardly necessary to add that the *R. nitidus* (Bell Salt.) is a very different plant which is called *R. Lindleianus* in this essay.

The *R. plicatus β carinatus* (Bell Salt.) seems to be nearly as closely, indeed probably more closely, allied to *R. affinis* than to *R. plicatus*. I incline to combine it with the former.

I possess a single specimen referred to above which was gathered near Bantry in the county of Cork, and closely resembles *R. montanus* (Wirtg.) but has characters more like those of *R. plicatus*. Both its shoots are thickly covered with short strong hooked prickles, and there are occasionally a few small prickles on the sepals. I believe it to be a form of *R. plicatus*. But a careful examination of the specimen of *R. montanus* (Wirtg.), published in the *Herb. Rub. rhen.* (No. 3.), leads me to concur in the opinion of Metsch (*Linnæa*, l. c. 140), that it is a form of *R. affinis*.

M. Genevier identifies a plant from Tory on Dartmoor with the *R. hamulosus* (Müll.). It has an abundance of strong declining (and some deflexed) prickles on its stem, and its panicle is furnished with rather numerous strong hooked prickles, but in other respects I do not see any characters by which to distinguish it from *R. plicatus*.

Mr Borrer remarks in his Herbarium that " Arrhenius shows *R. plicatus* (W. and N.) to be the primary *R. fruticosus* (Linn.), and a flowering specimen in *Herb. Linn.* (with authenticating number) confirms it. Some fragments of our *R. fruticosus* [*R. discolor*] are also preserved there and so named, but not numbered." It is nearly, if not quite certain that Arrhenius is correct, and that our present plant is the typical *R. fruticosus* (Linn.), although he certainly, as stated by Mr Borrer, included several other species under that name, especially in his later works. Wahlberg (*Fl. Goth.* 56) points out that the *R. maritimus* of the *Skanska Resa* (272) is our *R. plicatus;* and Linnæus quotes that plant as his *R. fruticosus* in the *Flora Suecica.* Wahlberg found it in the place mentioned by Linnæus. It was in the 2nd edition of the *Flora Suecica* that the confusion commenced, from Linnæus there adding some remarks which do not apply to *R. plicatus,* but to his *R. maximus fructu nigro* which is the *R. corylifolius* (Sm.). It seems to me that those authors act wisely who drop the name *R. fruticosus* as being only a cause of ambiguity. Specimens received from Sweden under the name of *R. fruticosus* are exactly our and the German *R. plicatus.*

If we are to judge from French specimens named *R. fruticosus v. intermedius* by Holandre the *R. Godronii* (Lec. et Lam.) is a state of *R. plicatus:* but if Godron's description is our guide we must place *R. Godronii* close to *R. corylifolius.*

The late Mr Bicheno, who paid much attention to the brambles, was satisfied that *R. plicatus* is distinct, for he said of his *R. ericetorum* from Snelsmore that it is " decidedly a good species." He never published his denomination of the plant, but specimens named by him are unquestionably *R. plicatus.*

Habitat.—Heathy places.　June, July.

Area.—1 2 3 . 5 6 . 8 9 10 11 12 13 14 15 16 . . 19 . . . 23 . 25 . . . . 30.

Localities.—i. Valley of the Tory, Dartmoor, *S. Dev.* (Briggs!).—ii. Seldown near Poole, *Dors.;* Burnt House, and America, *I. of W.;* Midhurst (Borr.), and St Leonard's Forest, *W. Suss.;* Forest Row, Frant, *E. Suss.*—iii. Walton, *Surr.;* Easney Park Wood, *Herts.;* Dartford, *W. Kent*

(Henslow!); Tonbridge Wells, *E. Kent* (Borr.!); Snelsmore
Common near Newbury, *Berks.* (Bicheno!); Snaresbrook,
S. Essex (E. Forster!).—v. Westfelton, *Salop.;* Baxterley
Common, and Bentley Wood, *Warw.* (Bloxam!).—vi. Falls
of Mynach, *Card.*—viii. Nailstone near Leicester, and
Twycross, *Leic.;* Chalk Abbey, *Derby* (Bloxam).—ix. Woolston Moss, and Paddington (W. Wilson!), *Ches.*; Agecroft
Hall near Pendleton (Herb. Wither.!), *S. Lanc.*—x. Wood
between Dalton and Sowerby near Thirsk, *N. E. York.*—
xi. Twizel House Dean, *North.*—xii. Rydal (Hort!), Haweswater (Borr.!), *Westm.;* Keswick, *Cumb.*

xiii. Jardine Hall, *Dumf.;* Stranraer, *Wigt.*—xiv. Black-
burn Rigg Dean, and banks of the Eye between Reston
and Covey-heugh Mill, *Berw.* (Johnston!).—xv. Killin, *Mid.
Perth.;* Between Stirling and Callander, *W. Perth* (Gre-
ville).—xvi. Arran, *Clyde Isles.*

xix. Bantry, *S. Cork.*—xxiii. *Meath* (D. Moore).—xxv.
W. Meath (D. Moore).—xxx. Kilrea, *Derry* (D. Moore).

N.B. It is possible that some of the localities from which
I do not possess specimens may be incorrectly given to *R.
plicatus*, for my views concerning it and *R. fissus* have
recently changed considerably. Berwickshire, W. Perth,
Meath, W. Meath, and Derry are thus doubtful.

6. **R. affinis** W. and N.

R. caule suberecto vel subarcuato angulato lævi gla-
briusculo, aculeis validis paululum deflexis declinatisve
è basi dilatata compressa conicis in angulos caulis congestis, foliis quinatis, *foliolis* coriaceis basi planis *apicem*
versus subundulatis utrinque viridibus opacis supra
subpilosis subtus pallidioribus sæpe sericeo pubescen-
tibus, foliolo terminali cordato-ovali cuspidato *infimis*
petiolatis *ramorum floriferorum basi attenuatis, pani-*
culæ compositæ foliosæ *ramis corymbosis* erecto-patentibus sæpe elongatis, sepalis acuminatis externe viridi-
tomentosis margine pallidioribus patentibus.

R. affinis Rubi Germ. 18. t. 3. Trattin. Ros. iii. 27.
Arrh.! 25.　Fries! Summa, 165;　Herb. Norm. vi. 45 (sp.).
Lees! in Steele, 59;　Malv. 57.　Leight.! in Phytol. iii. 73
(not of Fl. Shrop.).　Bab.! in A. N. H. Ser. 2. ii. 33;　Man.
ed. 3, 94; ed. 6, 106.　Blox. in Kirby, 47.　Sond. Fl.
Hamb. 273.　Fl. Dan. t. 2539.　Metsch in Linnæa, xxviii.
139.　Garke, Fl. Deutschl. ed. 7. 118.　Billot! Fl. Gall. et
Germ. exsic. No. 544 (sp.).　Wirtg.! Herb. Rub. No. 32 (sp.).
　R. fastigiatus Lindl.! Syn. ed. 1, 91.　　　　-
　R. incarnatus Müll. in Flora, 1858 (teste Genevier!).
　R. *fissus* Lees! MS.
　R. suberectus Lees! MS.
　R. plicatus β *racemosus* Lees! MS.
　Creeping. *Stem* suberect or elongate and arching but
perhaps never reaching the ground and rooting, nearly round
at the base, angular or even furrowed in the upper part,

purple, often slightly hairy. *Prickles* on the angles of the
stem strong, a little deflexed or declining, conical, from a
dilated compressed base; at the bottom of the stem they are
many, small, and patent. *Leaves* digitate-quinate, nearly
flat or slightly concave. *Leaflets* thin, dull green and distantly pilose above, rather paler with silky hairs or sometimes loosely felted beneath, flat at the base, wavy and a
little turned up at the rather irregularly or even doubly
serrate edges, especially towards the tip; basal leaflets
stalked, oblong, usually not over-lapping the intermediate
pair; intermediate broadly obovate-acuminate or nearly
round and cuspidate; terminal leaflet cordate-oval or cor-
date-orbicular, cuspidate; petioles (of which the common
are flat above but the partial furrowed) and midribs with
strong hooked prickles beneath; stipules linear-lanceolate.

Flowering shoot from whitely silky scales, angular, with
scattered hairs below. *Prickles* strong, deflexed from large
long compressed bases. *Leaves* nearly always ternate; floral
often simple. *Leaflets* broad, pilose above, more hairy and
paler beneath, irregularly and coarsely serrate, usually lobed
towards the tip; basal subsessile. *Panicle* hairy, branches
corymbose, erect-patent, hairy, branching only in their upper
part; or two or three of the lower axillary branches form
secondary racemose panicles; prickles strong, deflexed or
declining, the uppermost more slender. *Sepals* ovate, acuminate, greenish, aciculate, felted, hairy, with a strong edging of white felt, usually all patent with the fruit. *Petals*
rather broadly ovate, clawed, at first pinkish afterwards
white, slightly notched (Arrhenius and Metsch say that
they are quite entire). *Filaments* white or pinkish at their
base. *Anthers* and *styles* greenish. *Primordial fruit-stalk*
as long as the calyx. *Fruit* black.

Boreau says that the sepals are " étalés ou appliqués sur
le fruit à le maturité," Godron " appliqués," Metsch "patente

vix reflexo." They seem never to be very strongly reflexed, but rather patent or one or more sepals adpressed to the ripe fruit.

A plant called a weaker form of this species, var. *tenuis* (Blox. MSS.), mentioned in the *Flora of Leicestershire*, becomes strong and large in the Cambridge Botanical Garden. It differs from the typical state of the species by having patent prickles on both the stem and flowering shoot; the basal leaflets subsessile and lanceolate, intermediate lanceolate-acuminate, terminal leaflet obovate-lanceolate-acuminate; flowering shoot sometimes with quinate leaves like those of the stem, or when they are ternate the lateral leaflets are unequal-based and lobed. The flowers are racemose; the petals entire and white; the primordial fruit-stalk longer than the calyx. Mr Bloxam informs me that this is the plant which he formerly called (in letters to his correspondents) *R. Colemannianus.* The plant cultivated under that name in the Cambridge Garden, which was raised from seeds sent by Mr Bloxam, closely resembles it, but has rather scattered and rather unequal declining prickles, furrowed petioles, leaves nearly all ternate and flat, leaflets very broad and concave, primordial fruit-stalk much shorter than the calyx, and a stem which is rather arcuate-procumbent than erect-arcuate.

The *R. plicatus β carinatus* (Bell Salt.) is probably more correctly placed here than combined with *R. plicatus;* but its true position is rather doubtful. Its loose leafy panicle and the nearly glabrous back of its sepals resemble those of *R. plicatus;* but its elliptical leaflets are very dissimilar from those of any form of that species known to me. The lateral leaflets of the floral leaves, narrow gradually to their base as in *R. affinis.* This plant has received much less attention than it merits.

Many of the plants usually called *R. plicatus* belong more

correctly to *R. affinis;* but it must be confessed that we
now include under the latter name a somewhat heterogenc-
ous assemblage of forms, some of which will probably be
found to belong to other species when they are better known.
The true *R. affinis* (even the same bush) is very variable,
for its inflorescence forms a tolerably compound panicle or
quite simple raceme.

The *R. fastigiatus* of Lindley is considered by him to be
the same as his *R. fissus* (*Syn.* ed. 2); but the specimens
obtained from the Horticultural Garden with the former
name are *R. affinis* (W. and N.). One of them is stated to
have been brought from Dunkeld in Perthshire. Borrer
remarks that they are "like specimens from Mertens of *R.
affinis* (W. and N.)." The *R. fastigiatus* of Lindl. Herb.
is a remarkably large plant from Ayrshire. I believe it to
be *R. affinis,* although a detached leaf has its terminal
leaflet partially subdivided as in *R. fissus.* This leaf may
have been the cause of Lindley's opinion. I find no specimen of *R. fissus* in his Herbarium except one named *R.
suberectus var.* from Scotland.

The more markedly suberect forms of *R. affinis,* often
much resemble *R. plicatus;* they will be readily distinguished from it if attention is paid to the specific characters
of the species. The arching form approaches *R. rhamnifolius,* from which its different leaves and the felted border
of the sepals will distinguish it. *R. affinis* seems much
better placed amongst the *Suberecti* than with the *Rhamnifolii* where it has been arranged by British botanists.

The *R. affinis* of Leighton's *Flora* consists of only two
varieties, although their being marked as β and γ might
lead to the supposition that there was also a *var. a.* Of
these *var.* β is a form of *R. rhamnifolius;* but *var.* γ!
belongs, as stated by Leighton (*Phytol.* iii. 73), to *R. corylifolius.* The *R. affinis* of the 1st edition of my *Manual* is

also synonymous with *R. corylifolius*. The *R. affinis* of Leighton's *Fasciculus* is correctly named.

The plant called *R. affinis* by Mertens in Herb. Borr. agrees with our plant. The *R. affinis* of Billot (No. 544) is certainly the same plant as that known by the name in England, but the foliage of the stem is very imperfectly represented by the specimen contained in my copy of that collection: also its sepals are furnished with much more numerous, chiefly deflexed, aciculi than I have ever seen on the English plant. Sometimes indeed our plant is so sparingly furnished with aciculi that they might easily escape notice.

The *R. affinis* of Reichenbach's *Flora exsic.* (No. 781), collected by Weihe at Herford in Westphalia, has the leaves of the flowering shoot quinate or very nearly so, and the lower leaflets imbricate and rather enlarged at the base; also their under side is almost felted. It seems probable that Weihe or Reichenbach, has made a mistake and issued wrong specimens under this name. The plant in my set is apparently much more nearly allied to *R. corylifolius* than to the *R. affinis* of the *Rubi Germanici*.

β *lentiginosus;* caulis aculeis declinatis vel deflexis, foliolis subtus subglabris lanceolatis acuminatis serratis, foliolo terminali lanceolato basi paululum attenuato subcordato, paniculæ elongatæ foliosæ subsimplici aculeis uncinatis.

R. lentiginosus Lees! in Steele, 60 (1847).

R. fastigiatus Merc.! in Reuter Fl. Genev. 393.

R. incarnatus Müll. Mon. 22 (1859), teste Genevier.

This plant is probably a state of *R. affinis*, although Mr Lees still (*Bot. Worc.* 47) considers it distinct specifically. It seems to be almost exactly the *R. fastigiatus* of Mercier, but hardly of the *Rubi Germanici*. M. Questier sends a plant closely resembling it as *R. sylvaticus*, and M. Genevier names a similar plant *R. incarnatus* (Müll.). M. Müller places *R. sylvaticus* with his *R. incarnatus*.

Habitat.—Heaths and open woods. July, August.

Area.—1 2 3 . 5 6 7 8 9 10 . 12 13 14 15 16 . . 19.

Localities.—i. Hartknott wood near Ilfracombe, and Chambercombe, *N. Dev.;* near Plymouth, *S. Dev.*—ii. Bexley Heath, and Woodmancote (Borr.!), *W. Suss.*—iii. Between Cobham and Weybridge, *Surr.* (Borr.!); Tonbridge Wells, *W. Kent.;* Hatfield, *Herts.*—v. Stanton and Stapleton, *W. Glouc.;* Redwood near Cheltenham, *E. Glouc.* (Hort); Chepstow, *Monm.;* Malvern, *Worc.;* Chartley Moss, *Staff.;* Wrekin and Shawbury Heath, *Salop.*—vi. Fishguard and Milford, *Pemb.;* near Aberystwith, *Card.*—vii. Capel Curig, Llanberis and Bangor, *Caern.;* Pennal and Cwm Bychan, *Merion.;* Glan Hafren, *Montgom.*—viii. Stanton Harold, *Leic.;* Chalk Abbey, *Derby* (Blox.).—ix. Hale Moss near Bowdon (G. E. Hunt!), and Knutsford Moor, *Chesh.*—x. Hooton and Rotherham, *S. W. York.* (Blox.); Bilsdale, *N. E. York.*—xii. Keswick, *Cumb.;* Haweswater and Brathay, *Westm.*

xiii. Jardine Hall, *Dumf.;* Gouroch, *Renf.;* Wigton (Balf.).—xiv. Howgate, *Edinb.*—xv. Glen Falloch, *W. Perth.*—xvi. Lock Eil, *Western.;* Lamlash, *Arran* (Balf.).

xix. Killarney, *S. Kerry.*

GROUP II. RHAMNIFOLII.

Caules plus minusve arcuati, sparsim pilosi, nec pruinosi nec setosi neque tomentosi, radicantes. Aculei in caulis angulis sæpissime congesti, subæquales, in basi depressa compressa expansi.

The remarkably naked stems form the chief peculiarity of this group. They usually bear a few scattered hairs, and sometimes subsessile glands are found on the stem of *R. rhamnifolius* and *R. latifolius*. My acquaintance with two of the supposed species is very imperfect: one of them, *R. imbricatus*, certainly belongs to this group; whilst *R. latifolius* so much resembles *R. corylifolius* that it might possibly (but only possibly) have been better placed near to that species.

R. affinis is placed in the group *Suberecti*, owing to its much closer relationship to *R. plicatus* than to any of these plants. Its usual form is quite that of the *Suberecti*, but larger states closely resemble the *Rhamnifolii*. It seems to be the connecting link between the groups.

As it is certain that the original *R. nitidus* belongs to the *Suberecti*, we must change the denomination usually given to this group by English Botanists. It is now named from what seems to be the most prominent species included in it. It cannot be called *Corylifolii* with Lindley and Bell Salter; for *R. corylifolius* belongs to the group of *Cæsii*.

Dr Walker-Arnott remarks that the stems of these plants are not more without hairs and stellate down than those of the group *Villicaules*. But there is this difference between them: the stems of the *Rhamnifolii* are nearly naked even when very young, but those of the *Villicaules* become so only by age. The former plants also want the setæ, felted hairs, and aciculi which not unfrequently occur on the stems of the *Villicaules*.

7. R. Lindleianus Lees.

R. caule erecto-arcuato lævi nitido, aculeis validis
declinatis compressis basi dilatatis, foliis quinatis, foliolis subcoriaceis supra nitidis subtus pallide viridibus
pilosis (sæpe subtomentosis), foliolo terminali obovato
rotundatove acuminato infimis pedicellatis intermediis
dissitis, *paniculæ* compositæ foliosæ *ramis patentibus*
divaricatisve brevibus corymbosis *rachi polita in medio
spinosissima* superne pedicellisque tomentosis *aculeis
validis* declinatis.

R. *Lindleianus* Lees! in Phytol. iii. 361 (1848); Bot.
Malv. 57. Bab.! Man. ed. 5. 98; ed. 6. 107. Syme, Eng.
Bot. iii. 168.

R. *nitidus* Bell Salt.! in Phytol. ii. 101 (1845), (not of
Rub. Germ.). Bab.! Syn. 9; Man. ed. 2. 97; ed. 4. 96.
Leight.! in Phytol. iii. 75. Blox.! in Kirby, 46.

R. *leucostachys* Lindl.! Syn. ed. 2. 95 (1835), (not of
Sm. Eng. Fl.). Leight.! Fl. Shrop. 230. Lees! in Steele, 57.

R. *plicatus* Sm.! Eng. Fl. ii. 401 (1824), (not of Rub.
Germ.).

R. *rhamnifolius* β *nitidus* Bell Salt.! in Bot. Gaz. ii.
118; in Fl. Vect. 155.

R. *affinis* Sm. Eng. Fl. ii. 405 (1824),(not of Rub. Germ.).

R. *argeiracanthus* Müll. teste Genevier!

R. *hamulosus* Müll. Mon. 3 (1859), teste Baker!

Stem arching, angular throughout, not furrowed (except
perhaps at the extreme point), appearing as if varnished,
hairy near the base, with distant hairs in the upper part,
usually striate on the faces when young, but the striæ dis-

appearing with age. *Prickles* declining or subpatent, strong, compressed, dilated at their base. *Leaves* quinate. *Leaflets* subcoriaceous, flat or the edges turned up, wavy at the edge, irregularly or doubly dentate-serrate or with large coarse unequal-sided teeth which are themselves dentate-serrate, shining and with a few scattered hairs above, often so densely hairy beneath as to seem felted, obovate-lanceolate or even broader, shortly stalked; terminal broadly obovate-lanceoláte, acuminate; basal pair of leaflets much directed backwards in the plane of the leaf; furrowed petioles with many strong hooked prickles, especially at their upper ends; midribs slightly prickly; stipules slender.

Flowering shoot from ashy scales, hairy (the hairs usually adpressed), shining. *Prickles* slender, subpatent or much declining in the panicle. *Leaves* ternate. *Leaflets* obovate-lanceolate; lower divaricate, shortly stalked. *Panicle* leafy, compound, usually long, with a blunt convex end; branches many, short, patent, corymbose, dividing once or twice near to their top, few flowered; lateral peduncles longer than that of the terminal flower in each division of the panicle; rachis and peduncles very prickly at about their middle, but nearly unarmed at the base, hairy, with a very few sunken setæ or minute subsessile glands. *Sepals* ovate-acuminate, leaf-pointed, hairy, felted, with a very few sunken setæ and rarely an aciculus, reflexed. *Petals* not contiguous, oblong, clawed, slightly notched, white. *Filaments* white. *Anthers* and *styles* greenish. Primordial *fruit-stalk* usually short. *Fruit* small.

In the earlier part of the summer the stems often appear to be suberect, but as the season advances they extend so as to arch and usually reach the ground and root. The varnished surface of the stem and the polished cuticle of the rachis of the flowering shoot (which is seen notwithstanding its clothing of patent hairs) are very characteristic of this species, which has very many and very prickly divaricate

branches that are simple below, but again branch in a
divaricate manner at their top. Thus the whole inflorescence, in its most perfect state, forms a dense subcylin-
drical blunt panicle of which the ultimate subdivisions are
inextricably interlaced.

There is a form of this plant which has foliage very
nearly approaching the cordate-leaved state of *R. rhamnifolius*. It has much broader and less wavy leaflets than the
typical *R. Lindleianus:* the intermediate leaflets are rounded
at the base, and the terminal is more or less cordate below,
and usually broadest at or below its middle, and also rather
cuspidate than acuminate. But these leaflets vary in shape
upon the same bush. The leaflets are much more hairy and
often appear as if felted on the underside, or (on the plant
from Measham mentioned by Bloxam in the *Flora of Leicestershire*, and distributed by him in his *Fasciculus*) the hairs
have become so very numerous that the clothing is not to
be distinguished from felt. The stem of this plant from
Measham is similar to that of typical *R. Lindleianus;* the
panicle is more pyramidal, but similar in other respects.
One of the three specimens, derived from Leighton's Herbarium, which are marked as "*R. leucostachys*, determined
by Prof. Lindley," is this Rhamnifolius-like plant: the
others are typical *R. Lindleianus*. Another plant obtained
from the same Herbarium with the remark "*R. rhamnifolius*, la forme ordinaire" appended to it by Nees von
Esenbech, and called "typical *R. rhamnifolius*" by Borrer,
has the leaves of the same abnormal form but with few hairs
beneath, and also a panicle which is nearly typical.

I place here a plant noticed in Baker's *North Yorkshire*
(226) as a "peculiar small-leaved form" of this species, which
I possess from Gormire, where it is abundant, and which he
states to grow on "heathery ground in several places amongst
the eastern hills." It differs in a few points from the usual

state of the species. Its leaflets are slightly downy on the veins beneath, but otherwise glabrous; they are small and finely serrate so as much to resemble those of *R. carpinifolius:* often the double character of the dentition is not easily detected; although that is its structure on well developed leaflets. Its panicle exactly resembles that frequently seen on undoubted forms of *R. Lindleianus*, being small and open and comparatively few-flowered; but nevertheless possesses the structure characteristic of the species, although less finely and amply divided than that of the luxuriant plants to which Leighton applied the name of *R. leucostachys.* I have received an interesting specimen from Mr Lees of what seems to be a form of *R. Lindleianus.* Unfortunately I do not possess any part of its stem, but have one stem-leaf which is pinnate-septenate like those of the *Suberecti.* This confirms the idea that *R. Lindleianus* is closely allied to the *Suberecti;* but the inflorescence and calyx of *R. Lindleianus* are not like those of the plants of that group. This specimen grew on May Hill in Gloucestershire.

As the plant named *R. nitidus* by all continental botanists is very different from this species, and as some foreign authors still continue to think that their *R. nitidus* is distinct, although very closely allied to *R. plicatus;* it seems better to give up the name which was generally used for the present species in England. We therefore adopt the next oldest name, which happily is one which will commemorate the researches of a late eminent writer upon *Rubi,* viz. Dr John Lindley.

It is singular that Dr Bell Salter continued in his very last published remarks upon Brambles to identify his *R. nitidus* with that figured on tab. IV. of the *Rubi Germanici.* This is the more astonishing from his combining his *R. nitidus* with *R. rhamnifolius,* and placing it as the second variety of the species between *R. cordifolius* and *R. sylvati-*

cus (including *R. villicaulis*); for, if his *R. nitidus* is the
same as the plant so called on the continent, it belongs to a
different section of the genus. After the inspection of a
considerable number of authentic specimens it is clear to me
that Dr Bell Salter's *R. nitidus* does not include the *R.
nitidus* (W. & N.). Neither can I see any grounds for
combining it with *R. rhamnifolius* and *R. sylvaticus*, notwithstanding Dr Bell Salter's remark that they are "distinctly osculant." The plants seem to be very different and
each of them as constant to its characters as other well-marked
species. It might be supposed that my plant is not that
which Dr Bell Salter had in view, if we did not find that
in the *Botanical Gazette* (ii. 118) he distinctly quotes the
plant of my *Manual* as identical with his *R. nitidus*, and
that the several specimens named by him which are in my
possession are all *R. Lindleianus*. In that place he combines
R. affinis of Leighton and Babington with his *R. nitidus*.
It is my belief that all this confusion has resulted from the
original error of identifying our plant with the *R. nitidus*
(W. & N.). Dr Bell Salter seems never to have relieved
his mind from that mistake, notwithstanding the very conclusive paper by Mr Lees, which appeared in the *Phytologist*
in the year 1848. Some remarks upon specimens of *R.
rhamnifolius* β *nitidus* (Bell Salt.) will be found under *R.
rhamnifolius*.

It is remarkable that in my considerable collection of
continental *Rubi* there is no specimen agreeing with our *R.
Lindleianus*, nor have I been able to find any description in
foreign works which will apply to it.

Habitat.—Hedges and borders of thickets. July, August.
Area.—1 2 3 4 5 6 7 8 9 10 11 12 13 . 15 16.
Localities.—i. Thornbury, *S. Dev.*; Boniton Wood, *W.
Som.* (T. B. Flower).—ii. Ryde, *Isle of Wight.*—iii. Ditton
marsh and Walton, *Surr.*; Barrack wood, Warley, *Herts;*

Harrow Weald Common, *Middl.* (Hind!); Halsted and Snaresbrook (E. Forster!), *Essex;* Burnham, *Bucks* (Lees).— iv. *Northamp.* (Blox.); Fakenham, *W. Norf.* (Blox.).—v. May Hill and Stapleton, *W. Glouc.;* Broadheath and Malvern, *Worc.;* Ross, *Heref.;* Shrewsbury, *Salop;* Rugby and Ather- stone, *Warw.* (Blox.); Abergavenny, *Monm.* (Lees).—vi. New Radnor, *Radn.*—vii. Menai Bridge, Llanberis, and Capel Curig, *Caern.;* Capel Garmon, *Denb.;* Pennal and Dolgelly, *Merion.; Anglesea* (W. Wilson!).—viii. Thring- stone and Twycross, *Leic.; Derby* (Blox.).—ix. Hale Moss near Bowdon (G. E. Hunt!) and Knutsford, *Chesh.*—x. Thirsk and Scarborough, *N. E. York.;* Bell Hag near Sheffield, *S. W. York.*—xi. Barnard Castle, *Durh.*—xii. Keswick, *Cumb.;* between Furness and Rampside, *N. Lanc.;* Stock Gill near Ambleside (Borr.!), Bowness (Hailstone!) *West.;* Douglas, *Isle of Man.*

xiii. Gourock, *Renf.*—xv. Alva, *Clackm.* (Balf.!).—xvi. Arran, *Clyde Isles* (Balf.); Tarbet, *Dumb.* (Hailstone!)   •

8. R. rhamnifolius W. and N.

R. caule arcuato angulato supernè sulcato, aculeis validis patentibus declinatisvc, foliis quinatis, *foliolis* coriaceis *planis* supra opacis subtus viridi-albo-tomentosis, foliolo terminali obovato vel cordato subcuspidato *infimis* petiolatis *intermediis dissitis,* paniculæ tomentosæ sæpe ad apicem densæ obtusæ ramis axillaribus racemosis paucifloris distantibus aculeis validis declinatis.

R. rhamnifolius Rubi Germ.! 22. t. 6 (1822). Sm. Eng. Fl. ii. 401. Borr. in Eng. Bot. Suppl. t. 2604; in Hooker, ed. 2. 244; ed. 3. 248. Bab.! Man. ed. 1. 93; ed. 6. 107. Leight.! Fl. Shrop. 227 (first form in part). Billot! Fl. Gall. et Germ. exsic. No. 543 sp.). Syme, Eng. Bot. iii. 186. t. 446.

R. cordifolius Rubi Germ. 21. t. 5 (1822). Bab. Syn. 13; Man. ed. 2. 98. Lees! in Steele, 59; Malv. 56. Leight.! in Phytol. iii. 173. Blox.! in Kirby, 49. Bor. Fl. Centre, 203.

R. rhamnifolius a cordifolius, and γ *sylvaticus* Bell Salt.! in Fl. Vect. 155.

R. affinis β Leight.! Fl. Shrop. 226.

R. Thuillieri Poir. Dict. Suppl. iv. 694 (1816)? Bor. Fl. Centre, 203?

R. thyrsoideus β *rhamnifolius* et γ *cordifolius* Bluff et Fingerh. ed. 2. i. pt. 2. 192. Metsch in Linnæa, xxviii. 126.

R. argentatus Müll.! in Jahresb. Pollichia, xvi. 93.

Stem angular below with flat sides, usually furrowed towards the top, nearly glabrous, but with a few small scattered hairs and subsessile glands, usually bright red and

shining. *Prickles* strong, straight, patent or declining, from
a long compressed base, usually yellow or tipped with red or
tinged similarly with the stem. *Leaves* quinate. *Leaflets*
coriaceous, flat, usually finely but not quite equally serrate,
or slightly doubly serrate towards the tip, dark dull green
with a few hairs chiefly on the ribs above, hard grey- or
greenish-felted and with hairs on the ribs beneath; all stalked;
basal with manifest but short stalks, obovate or oblong,
acuminate, not imbricate, usually spreading or directed
backwards; intermediate long-stalked, obovate, cuspidate;
terminal long-stalked, often slightly cordate, and broad at
the base, then widening gradually up to about the middle,
from thence narrowed to an acute or sub-cuspidate point;
furrowed petioles and under side of *midribs* with strong
hooked prickles; stipules linear-lanceolate.

Flowering shoot from brown scales clothed with ashy
down, angular or slightly furrowed, with clustered hairs and
subsessile glands. *Prickles* few, strong, deflexed, from a very
long base, reddish with yellow tips. *Leaves* ternate, rarely
quinate, the uppermost floral leaves simple. *Leaflets* clothed
like those of the stem, obovate or oblong, acuminate or cuspidate, often incise-serrate, lower unequal at the base; when
quinate the basal leaflets are sessile and, as well as the intermediate, are usually wedge-shaped below; petioles and under
side of midribs with small hooked prickles. *Panicle* sometimes rather pyramidal, usually broad, blunt, convex and dense
at the top, compound, hairy, ashy-tomentose with very short
setæ especially towards the top of the rachis and branches;
prickles many, long, slender, deflexed, from very long bases;
branches axillary, patent or ascending, racemose, rather
distant, becoming shorter, closer together, and more corymbose
upwards, upper branches from the axils of simple ovate
leaves, few uppermost extra-axillary. *Sepals* ovate-acuminate with a narrow point, hairy and felted, and with a few

prickles at the base externally, reflexed. *Petals* roundish, slightly wavy, clawed, white. *"Filaments* white. *Anthers* and *styles* pale green." Primordial *fruit-stalk* shorter than the *sepals*, which are loosely reflexed and point downwards. *Primordial fruit* oblong. *Seed* broadly half-ovate, blunt; inner edge nearly straight; sides convex.

The form of leaflet described above may perhaps be considered as typical, but it is far from being constant; for plants may be found having a nearly ovate or even round leaflet, which is cuspidate, rather than acuminate.

The *R. rhamnifolius* of Leighton's *Flora* includes two plants. His "first form" is that which is considered as the true plant by all authors who have noticed it; but he combined with it a plant having a "very hairy [stem] with numerous minute glands interspersed"(!), which appears to me to be perhaps more correctly placed under *R. carpinifolius*. Leighton's "second form" was named *R. corylifolius* by Borrer and *R. rhamnifolius* by Nees and also by Lindley. The former seems to be the more correct view; and therefore it will be found noticed below as *R. corylifolius* γ *purpureus*.

Our *R. rhamnifolius* is probably the *R. thyrsoideus* γ *rhamnifolius* of Bluff and Fingerhuth, Sonder, and Metsch. They show only slight reason for combining *R. thyrsoideus* and *R. rhamnifolius*, and I am unable to agree with them. To me the plants seem even more different in reality than they can be shown to be by description. Godron combined them in his *Monographie* (1843), but separates them in the *Flore de France* (1848), and in the second edition of his *Flore de Lorraine* (1857). He points out that the petals of *R. rhamnifolius* are very round, and not narrowed to the base but clawed; the stem has flat sides except in its superior part: also that the petals of *R. thyrsoideus* are obovate and narrowed to their base; and the stem furrowed throughout. It must be admitted that the stem of *R. rhamnifolius*, as represented in

Rubi Germanici, is very unlike that of our plant, which closely resembles what is figured in that work as *R. cordifolius*. The panicle there given as that of *R. rhamnifolius* well represents that of our *R. rhamnifolius*, although with us the upper part is often shorter, denser, and more dome-like. The specimen in my copy of Leighton's *Fasciculus* is very characteristic and agrees exactly with that called "R. rhamnifolius, la forme ordinaire" by Esenbech. I do not find more than one specimen that agrees well with our *R. rhamnifolius* in my rather large collection of foreign Rubi. I identify our plants with the *R. rhamnifolius* and *R. cordifolius* of the *Rubi Germanici* on account of their agreement in most respects with the plates and descriptions in that work and with the specimen named by Nees von Esenbech for Leighton. Nevertheless it seems not impossible that our plant may really be different from that similarly named by continental botanists. That is a question which I have found myself unable to decide without the aid of good and authentic foreign specimens. It must be left for determination by some botanist more fortunately situated in that respect.

The *R. rhamnifolius* of Billot seems to agree very well with the typical plant (as figured in *Rubi Germ.*) and with English specimens.

As has already been remarked, the fully developed panicle of *R. rhamnifolius* has a rather pyramidal outline, but it is very blunt and dense at the top. As the distance from the top increases the branches lengthen and separate more and more from each other; but even the lowest branch falls short of its accompanying leaf. Nevertheless occasionally the panicle is narrower and somewhat thyrsoid in its upper part, although even then it is blunt. Cordate or ovate or obovate terminal leaflets seem to accompany either of these forms of panicle indifferently; but perhaps the narrower panicle is most usually the produce of plants having the cordate leaflet. Apparently the converse is the fact in Germany, if I am correct in identifying our plants with those of the *Rubi Germanici*. All the leaflets of our cordate-leafed plant are (upon both the stem and flowering shoot) shorter and broader than those of our typical *R. rhamnifolius;* those of the stem tend towards a cordate form, especially at their base, and are usually dentate; the basal leaflets are usually broadly ovate or oblong, the intermediate broadly obovate, the terminal roundly or sometimes almost exactly cordate but with a cuspidate point. There seems to be no true distinction between the plants.

The presence of felt on the leaves is undoubtedly (in my opinion) typical of *R. rhamnifolius;* yet plants may be found having much hair, and very little or no felt on those organs. Their leaves are also more dentate than is usual in this species, with which however it seems proper to place them.

Bell Salter combined *R. nitidus, R. affinis, R. sylvaticus* and *R. villicaulis* with *R. rhamnifolius.* The two former seem to me, and to most other students of Brambles, to be abundantly distinct. The two latter are so different in most respects that it is interesting to find an apparent cause for what I cannot but consider as a great mistake. Dr Salter mentions only one station for his *R. sylvaticus* (including *R. villicaulis*), viz. "In a hedge at Weeks-field near Ryde," in the Isle of Wight. Fortunately I possess a specimen from that place, gathered, named *R. villicaulis,* and given to me by Dr Bell Salter. Its stem is not in the least "villose," even in the young state which my specimen represents, but bears only a few scattered hairs. I name it *R. rhamnifolius.* Dr Bell Salter's *R. villicaulis* is therefore not the true plant so named, but simply a form of *R. rhamnifolius.*

I am also able to determine with tolerable certainty the

plant intended by Dr Bell Salter when quoting *R. affinis* as a synonyme of his *R. rhamnifolius β nitidus* (*Bot. Gaz.* ii. 118), for there are three specimens named *R. affinis* in his Herbarium, from the Isle of Wight, from Selborne, and from Poole. The Isle of Wight plant and that from Poole are *R. corylifolius β conjungens* (Bab.) : that from Selborne, which is noticed by him in the *Phytologist* (ii. 100), is imperfect, but is certainly neither *R. affinis* nor *R. corylifolius*. I quite believe it to be the plant which I now call *R. althæifolius*.

Prof. Boreau changes the name of this plant to *R. Thuillieri* (Poir.). It does not seem desirable to alter a well-known and now universally adopted name because we fancy, for the proof seems to be very imperfect, that this is the plant called *R. tomentosus* by Thuillier and *R. Thuillieri* by Poiret.

Sonder (*Fl. Hamb.* 275) says that the *R. rhamnifolius* of *English Botany* (*Suppl.* t. 2604) is not a form of his *R. thyrsoideus*, to which he refers the *R. rhamnifolius* and *R. cordifolius* of the *Rubi Germ.*, on the authority of specimens named by Weihe. I have often suspected that the two authors of that great work did not always concur in their nomenclature when naming or distributing specimens. The typical specimens that Leighton obtained were all named by Nees von Esenbech, most of those quoted in the continental books were from Weihe. Sonder adds that a specimen received from me as *R. rhamnifolius* must be named *R. discolor*. I fear that this shows carelessness on my part, or perhaps ignorance of the true plant at the time (many years since) when the specimens were sent.

M. Genevier states that Mr Briggs's Devonshire specimens are the *R. argentatus* (Müll.), and says "cette plante est très éloignée du *R. rhamnifolius* (W. and N.)." An examination of the specimens had previously led me to the same conclusion, except that I consider them to be undistinguishable from *R. rhamnifolius.*

Habitat.—Hedges and thickets. July, August.

Area.—1 2 3 4 5 . 7 8 . 10 . 12

Localities.—i. Saltash, *S. Dev.* (Briggs!).—ii. *Isle of Wight;* Woodmancote and St Leonard's Forest, *E. Suss.* (Borr.!).—iii. Messing, *N. Essex* (Varenne!); Speen, *Surr.* —iv. Lynn, *W. Norf.*—v. Forest of Dean, *W. Glouc.;* Trellech and Llanrumney, *Monm.;* Broadheath, *Worc.;* Shrewsbury, *Salop;* Ross, *Heref.* (Purchas!).—vii. Llanberis, *Caern.* —viii. Twycross, *Leic.*—x. Thirsk, *N. E. York.*—xii. Ambleside, *Westm.;* Douglas, *Isle of Man.*

9. R. incurvatus Bab.

R. caule arcuato-prostrato angulato, aculeis validis patentibus declinatisve, foliis quinatis concavis, *foliolis* coriaceis *marginem versus incurvatis* undulatisque acuminatis supra nitidis et subglabris *subtus viridi-albotomentosis,* foliolo terminali cordato-ovato, paniculæ angustæ inferne foliosæ *ramis* brevibus *corymbosis* patentibus approximatis apice et pedicellis hirtis *tomentosisque* aculeis validis tenuibus deflexis, sepalis ovato-acuminatis.

R. incurvatus Bab.! in A. N. H. Ser. 2. ii. 36 (1848); Man. ed. 3. 95; ed. 6. 107. Syme, Eng. Bot. iii. 169. Lees, Malv. 55?

Stem arcuate-prostrate, slightly angular and very hairy at the base, slightly hairy throughout, angular, furrowed. *Prickles* strong, straight, declining, from a long compressed base. *Leaves* quinate. *Leaflets* coriaceous, flat except at the edges which are wavy and turned upwards, i.e. towards the upper side of the leaf, doubly dentate, shining above, soft hairy-felted greenish white beneath; all stalked, acuminate; basal .very shortly stalked, oblong, sometimes overlapping the intermediate pair; intermediate oblong-obovate; terminal roundly cordate-obovate; petioles flat above or very slightly furrowed, and together with the midribs having strong hooked prickles beneath; stipules linear-lanceolate.

Flowering shoot rather long, from white scales clothed with silky down, patently hairy. *Prickles* few, strong, short, deflexed. *Leaves* ternate, uppermost floral leaves simple. *Leaflets* pilose above, pale green and hairy beneath, nearly

equal, obovate or oblong; lateral lobed externally; petioles
and under side of midribs with small hooked prickles.
Panicle narrow, felted, pilose, with short yellow sunken
setæ; prickles long, declining, or a little deflexed, rather
slender; branches short, patent, corymbose, 2 or 3 lowest
axillary and distant, upper close together; upper two thirds
of the panicle leafless; sometimes the lowest branch forms
a secondary panicle. *Sepals* ovate, acuminate, leaf-pointed,
hairy and felted externally, reflexed from the fruit but their
points turning upwards. *Petals* roundly obovate narrow to
their base, pink, finely serrate. *Filaments* pink at the base.
Anthers yellowish. *Styles* pinkish at the base. *Primordial
fruit-stalk* shorter than the sepals. *Primordial fruit* hardly
more than hemispherical. *Seed* ovate, very broad at the
base; inner edge nearly straight.

The wavy edges of the leaflets turning upwards distinguish this plant from all its allies. Each leaflet is
concave: in *R. imbricatus* it is "convex from the tendency
of the edges to turn downwards." The panicle of *R. incurvatus* has shorter and more closely-placed branches, and is
therefore closer than that of *R. imbricatus*. The points of
the sepals are not directed downwards as in *R. rhamnifolius;*
but, although strongly reflexed at their base, form a continuous curve, so as to direct their points upwards. The
petals are not clawed, but narrow gradually.

The specimen gathered in the Isle of Man has a much
more leafy panicle than is usual but agrees with this species
in other respects. It is an abundant plant there.

The plant from Lyth Hill near Shrewsbury, mentioned
in the *Annals of Natural History* (l. c. 38), is possibly an
anomalous state of this species. Its leaves are in an unnatural condition and appear to be without felt; also the
basal and intermediate leaflets have longer stalks than is
usual. It may belong to *R. rhamnifolius.*

I have not seen any continental specimens which agree with this plant, but as it is very abundant in the valley of Llanberis, and is reproduced from seed, I have much confidence in its distinctness as a species.

Habitat.—Heaths and open woods. July.

Area.— . 2 3 . . 6 7 . . . . 12 . . . 16.

Localities.—ii. Rotherbridge, *W. Suss.*—iii. Richmond, *Surr.*—vi. Milford, *Pemb.*—vii. Llanberis abundantly, Capel Curig and Bangor, *Caern. ;* Pennal, Dolgelly and Cwm Bychan, *Merion.*—xii. Douglas, *Isle of Man.*

xvi. Dunoon, *Renf.* (Balfour!).

10. R. imbricatus Hort.

R. caule arcuato-prostrato ramosissimo angulato, aculeis parvis validis è basi valde dilatata compressa declinatis, *foliis convexis* quinatis, *foliolis convexis* coriaceis supra opacis et subglabris subtus pallidioribus sparsim pilosis *imbricatis cuspidatis,* foliolo terminali rotundo-obovato-cordato, *paniculæ* angustæ infernè foliosæ *ramis longis racemosis* ascendentibus distantibus apice et *pedicellis hirtis vix tomentosis* aculeis brevibus tenuibus deflexis, sepalis abrupte cuspidatis.

R. imbricatus Hort! in A. N. H. Ser. 2. vii. 374 (1851). Bab.! Man. ed. 3. 94; ed. 6. 108. Syme, Eng. Bot. iii. 170.

Stem arcuate-prostrate, with many 'slender whiplike shoots, angular, slightly furrowed, purplish red, nearly or quite glabrous. *Prickles* slender but strong, declining, from a much dilated compressed base. *Leaves* quinate. *Leaflets* convex, slightly wavy throughout, opaque and pilose above, paler and pilose beneath, doubly but not deeply dentate-serrate, basal overlapping the intermediate which overlap the terminal leaflet; basal oblong, cuspidate, shortly stalked; intermediate obovate, cuspidate; terminal roundish-obovate with a cordate base; petioles all flat above, or the partial ones channelled, bearing together with the midribs strong decurved prickles beneath; stipules linear.

Flowering shoot from brown scales clothed with whitish silky hair, nearly glabrous, but having a few patent hairs. *Prickles* small, strong, declining. *Leaves* quinate or ternate. *Leaflets* subglabrous above, paler and pilose beneath, cordate-ovate or cordate-obovate; petioles and midribs with very slender deflexed prickles beneath. *Panicle* rather narrow, slightly hairy below, very hairy but scarcely, if at all, felted at the top, but the very top of the panicle and the peduncles are furnished with a thin coat of stellate hairs, amongst which are many sunken setæ; floral leaves often simple, cordate, somewhat three-lobed; prickles few, short, slender, declining, from a long compressed base; branches falling short of the leaves, racemose, ascending, 3 or 4 lowest axillary and distant, uppermost subcorymbose or even single-flowered. *Sepals* ovate, abruptly cuspidate, with a short linear point, or narrowly leaf-pointed, clothed with ashy felt and having an occasional minute prickle. *Petals* obovate, narrowed to their base, white, notched at the end. *Styles* greenish yellow below. *Primordial fruit-stalk* longer than the calyx. *Primordial fruit* rather small, subglobose, glossy black.

The Rev. F. J. A. Hort has taken great pains to distinguish this plant from its allies, and as an isolated paper is not unlikely to be overlooked, it is desirable to transfer a portion of his remarks to this place. He says "It is closely allied to *R. affinis, R. cordifolius* [*R. rhamnifolius*], and *R. incurvatus.* On a hasty inspection it might probably be referred to *R. corylifolius*, but there is in reality a wide gap between them, the latter species being rightly referred to the group of *Cæsii.* It is often difficult to distinguish dried specimens of *R. imbricatus* and the three species above mentioned, although no one accustomed to Brambles could confound them when growing. The present plant may be known from the larger and more typical forms of the protean *R. affinis* by the structure of the branches of the panicle, which are racemose and not cymose, and their much slighter degree of divarication from the rachis, and by the sepals being abruptly cuspidate and not gradually acuminate; (to the less developed forms which apparently constitute Mr Lees's *R. lentiginosus*, having suberect stems and nearly simple panicles and growing chiefly in heathy places, it bears no resemblance): from *R. cordifolius* [*R. rhamnifolius*] by the laxer and less pyramidal panicle, the absence of tomentum [felt] on the underside of the leaves, and the agreeable flavour, globular shape, and glossy lustre of the fruit, which in the latter species are very peculiar, when able to ripen freely, being remarkably large, oblong, with somewhat flattened drupes, dull and burnished rather than glossy, and very insipid (it should be observed that all these three species grow in the same neighbourhood): from *R. incurvatus* by the leaves being hairy but not covered with a firm velvet beneath, and by the yellowish-green not flesh-coloured styles. The numerous secondary shoots of the barren stem, the imbricated and convex leaves and leaflets, and the absence of tomentum on the upper part of the panicle, sufficiently separate it from all three species." *Ann Nat. Hist.* l. c. 375—376.

I have very little acquaintance with this plant, never having seen it growing, but have great confidence in the accuracy and judgment of its describer. The convex state of the leaves and also of the leaflets must cause it to differ remarkably in appearance from *R. incurvatus*, which the dried and pressed specimens greatly resemble. It is also much like some states of *R. corylifolius;* but wants the bloom, the smaller scattered prickles, and the more or less plentiful setæ of that species. I am unable to identify it with any described plant, and recommend it to the study of botanists visiting the beautiful district which it inhabits.

Habitat.—Thickets. June, July.

Area.— . . . . 5.

Localities.—v. "In many places mostly on sloping banks, for three or four miles on both sides of the Wye below Monmouth, in both Monmouthshire and Gloucestershire: especially by the tramroad above Redbrook." *Hort.*

11. R. latifolius Bab.

R. caule arcuato-prostrato angulato *sulcato*, aculeis parvis tenuibus compressis e basi longissima compressa subdeclinatis, foliis quinatis, foliolis utrinque pilosis grande- et duplicato- dentatis tenuibus subtus nunquam tomentosis, *foliolo terminali cordato-acuminato* infimis sessilibus intermediis incumbentibus, paniculæ brevis foliosæ pilosæ ramis ascendentibus paucifloris corymbosis apice et pedicellis tomentosis hirtis aculeis brevibus tenuibus declinatis.

R. latifolius Bab.! Man. ed. 3. 94 (1851); ed. 6. 108; in A. N. H. Ser. 2. ix. 124.

R. Wahlbergii Lange, Danske Fl. 350?

Stem usually quite prostrate, angular and furrowed throughout, nearly glabrous but with scattered subsessile glands, not stellately downy nor setose. *Prickles* nearly all placed on the angles of the stem, rather few, moderately long, slender from a long compressed base, straight, declining, nearly equal; rarely one very much smaller may be found. *Leaves* quinate. *Leaflets* very broad and large, dull green and pilose above, paler and with numerous hairs beneath, coarsely and irregularly doubly dentate; basal broadly oblong, rather rhomboidal, sessile, overlapping the intermediate pair which are of similar shape but larger and shortly stalked; terminate leaflet with a stalk equalling one-third of its length, cordate-acuminate; *petioles* furrowed above; and as well as the midribs yellowish and with a few small weak declining or slightly deflexed prickles beneath; *stipules* leaflike, lanceolate-attenuate.

Flowering shoot long, surrounded at its base by short scales ashy with silky pubescence, angular, green, nearly glabrous. *Prickles* few, short, weak, from long bases, slender, declining, yellow tinged with purple. *Leaves* ternate. *Leaflets* pilose on both sides but chiefly beneath, nearly equal, ovate, acute, deeply and doubly serrate, lower ones often strongly lobed on the outer edge below; petioles with very few slender declining or deflexed prickles; midrib unarmed or with small prickles; *stipules* linear-lanceolate. *Panicle* short, leafy below, pilose; the upper part and pedicels felted, pilose, and with a few short sunken setæ or subsessile glands; prickles short, declining, slender, yellow; branches short, ascending, few-flowered, corymbose; bracts trifid with narrow lanceolate segments. *Sepals* ovate-acuminate, felted on both sides, whitish within, rather green and pilose externally, reflexed loosely from the fruit. *Petals* shortly ovate, clawed. *Primordial fruit* apparently hardly more than hemispherical. The flowers and fruit require more careful examination.

The leaflets have very large acute teeth, almost amounting to lobes on both of my Scottish plants (but on that from Monmouthshire, although they are coarsely and doubly dentate, the large double teeth are not so conspicuous); the teeth or lobes are themselves irregularly and acutely toothed, and are divided from each other by very acute angles. The leaves are truly dentate, although the teeth are all slightly directed forwards; none of the teeth are patent nor divaricate. The Monmouthshire specimen, for I unfortunately gathered only one, has rather stronger prickles, its terminal leaflets roundly cordate-obovate and cuspidate-acuminate, its basal leaflets very shortly stalked; but in other respects it accords with the Scottish plants.

In my *Synopsis* I placed this plant doubtfully under *R. Salteri*, to which I am now convinced that it has no relationship. It much resembles some states of *R. corylifolius* in foliage; but differs by its deeply furrowed stem, want of felt on its leaves, prickles confined to the angles of the stem, and the total absence of aciculi and setæ. The panicle also wants the long spreading branches which are usually conspicuous in *R. corylifolius.*

Mr Kirk justly remarks (*Phytol.* iv. 969) that the plant found near Thirsk to which I once gave the name of *R. latifolius* is not distinguishable from *R. corylifolius* β *conjungens.* I have only seen one specimen of it. He says that it sometimes has a furrowed stem, which is unusual in *R. corylifolius.* Its leaves are very different from those of the true *R. latifolius.*

Mr Lange quotes a specimen named *R. latifolius* by me, as belonging to the *R. Wahlbergii* (Arrh.). As I do not know from whence the specimen was obtained, it is possible that it may be the misnamed plant from Yorkshire. In my opinion the *R. Wahlbergii* is not distinguishable from my *R. corylifolius* β *conjungens.*

The fact that this plant has been noticed in only three or four places and in very small quantity renders it probable that it is an abnormal state of some better understood species. Its fruit is unknown. If a distinct species its true place in the genus is not yet determined. It associates very badly with the *Rhamnifolii*, although agreeing with them in technical characters.

Habitat.—Open woods. July, August.

Area.— . . . . 5 . . . . 14 15.

Localities.—v. By the tramway near Lower Redbrook near *Monmouth.*

xiv. By the river above Cramond Bridge near Edinburgh, *Linlithg.;* at Colinton near Edinburgh, *Edinb.* (Balfour!).— xv. In a wood below the road from Kenmore to Acharn, *Mid Perth.*

GROUP III. VILLICAULES.

Caules plus minusve arcuati, radicantes, pilosi vel
calvati, sæpe tomentosi, glandulis subsessilibus; vel
raro setosi aciculatique. Aculei in caulis angulis congesti, subæquales; vel etiam paucis minoribus sparcis.
Foliola infima petiolata intermediis dissita (*R. Gra-
bowskio* excepto).

There are two ways in which the *Rubi Villicaules* may
be divided into minor groups. If the direction of the stem
is alone considered we have (1) those plants in which it is
erect-arcuate, often never reaching the ground so as to root
at the end, or only doing so by means of a slender nearly
leafless autumnal shoot. These stems are usually very erect
and strong enough to support themselves in an upright
position. Such plants are *R. carpinifolius* and *R. thyrsoideus.*—(2) The stems are truly arcuate and nearly always
reach the ground and root (not requiring a special autumnal
shoot to do so), but rarely have much if any prostrate portion
at the end. The arch is lofty and self-supporting. Here
we may place *R. Grabowskii*, *R. villicaulis* and *R. mucronulatus.*—(3) The other plants included in the group have
arcuate-prostrate stems when left without any foreign support, and the prostrate part is usually very long relatively
to the low arch formed near the base of the stem. But this
mode of subdivision is far from being satisfactory. It
separates to a long distance from each other some very
closely allied plants and places together others which have
not much in common.

9

Another mode of forming three subordinate groups is pointed out by P. J. Müller, which, although less simple, is more natural. His groups are characterised at considerable length in his Monograph, but their more marked distinctions seem to be as follows.

1. *Discolores.*—Stem bearing equal strong prickles and adpressed pubescence. Leaves white-felted beneath.—*R. discolor, R. thyrsoideus.*

2. *Sylvatici.*—Stem bearing equal moderate (in size and strength) prickles and patent dense hairs. Leaves green, rarely white-felted beneath.—*R. leucostachys. R. Grabowskii. R. Salteri. R. carpinifolius. R. villicaulis. R. macrophyllus.*

3. *Spectabiles.*—Stem bearing more or less unequal prickles, a few scattered aciculi and often a very few setæ, also often densely pubescent.—*R. mucronulatus. R. Sprengelii.*

The first of these groups seems quite separable from the others, if our plants alone are considered. The second and third graduate into each other: for the typical state of *R. leucostachys* belongs to the *Sylvatici*, but the *R. vestitus* would better range amongst the *Spectabiles*, and yet there can be no doubt of their constituting only one species. Similarly the original *R. Salteri* belongs to the former group, and *R. calvatus*, which I do not distinguish specifically from it, is one of the *Spectabiles*. This third group also is most closely connected with the *Glandulosi*; for the *Radulæ*, which it seems probable that Mr Müller includes amongst his *Spectabiles*, for he places *R. rudis* there, form a well marked group of species connecting the glandular brambles with those whose stems are devoid of stalked glands (setæ). They may be shortly characterised as follows:

4. *Radulæ.*—Stem bearing nearly equal prickles, and also many short, nearly equal and deciduous, aciculi and

setæ seated upon minute tubercles which render the old stems rough like a file.

All the *Villicaules* are liable to have their stems denuded (calvati) when full grown, and then they are sometimes difficult to distinguish from the *Rhamnifolii:* but if the younger states of the stem are examined it is believed that the characteristic covering will be always found present in more or less abundance. The stems of the *Rhamnifolii* seem never to have setæ, nor to be felted, even in their youngest state.

a Discolores.—Aculei caulis æquales, validi; pubescentia arcte adpressa. Folia subtus cano-tomentosa.

12. R. discolor W. and N.

R. *caule arcuato-prostrato* angulato sulcato *stellato-sericeo* (griseo), aculeis e basi valde dilatata compressa declinatis vel deflexis, foliis quinatis, *foliolis* convexis coriaceis supra rugulosis *subtus tenuissime cano-tomentosis*, foliolo terminali obovato-cuspidato, paniculæ elongatæ contractæ tomentosæ ramis inferioribus axillaribus paucis multifloris aculeis validis uncinatis, *calyce tenuissime cano-tomentoso*.

R. discolor Rubi Germ.! 46. t. 20 (182⁴⁄₅). Reichenb.! Fl. excurs. 603; Fl. exsic. 1058 (sp.). Arrh.! in Fries Nov. Mant. iii. 40. Fries! Summa, 165; Herb. Norm. viii. 48 (sp.). Sond. 277. Bab.! Man. ed. 2. 99 (excl. var. β et δ); ed. 6. 108. Leight.! Fl. Shrop. 228; in Phytol. iii. 174 (excl. var. γ et δ). Lees! Malv. 57. Bell Salt.! in Bot. Gaz. ii. 121 (excl. var. β); in Bromf. Fl. Vect. 157 (excl. var. β). Bor. Fl. Centre, ii. 198. Metsch in Linnæa, xxviii. 151. Drejer! Fl. Hafn. 181. Billot! Fl. Gall. et Germ. exsic. No. 1659 (sp.). Syme Eng. Bot. iii. 171. t. 447. Merc. in Reut. Cat. Genev. 278.

R. fruticosus Sm.! Fl. Br. ii. 543 (1800); Eng. Bot. t. 715; Eng. Fl. ii. 399 (in part). Lindl.! ed. 1. 92; ed. 2. 95. Bab. Prim. Fl. Sarn. 31; Man. ed. 1. 94. Blox.! in Kirby, 45. Borr.! in Hook. ed. 2. 245; ed. 3. 248. Leight.! Fl. Shrop. 229. Lees in Steele, 58 (excl. var. γ et ε).

R. abruptus Lindl.! ed. 1. 92.

R. discolor—argenteus Bell Salt. in A. N. H. xvi. 367. Leight.! in Phytol. iii. 175.

R. discolor—macroacanthus Bell Salt.! in A. N. H. xvi. 366.

R. rusticanus Merc.! in Reut. Cat. Genev. 279.

R. Bastardianus Genev. in Obs. Rub. Herb. Bastard, 10.

Stem nearly prostrate unless supported, nearly round at the base with a few patent unequal prickles and scattered hairs, often covered with a fine glaucous bloom, soon becoming angular, furrowed near the top, bearing many minute stellate hairs. *Prickles* large, strong, patent, compressed, from a dilated base, seated on the angles of the stem. *Leaves* quinate and ternate on the same plant. *Leaflets* hairy beneath towards the base of the stem, others closely white-felted beneath, all stalked, rather finely and often doubly dentate-serrate, very variable in form, usually the lower leaflets ovate-lanceolate; intermediate and terminal obovate, acute, but sometimes (*R. abruptus* Lindl.!) cuneate-oblong abruptly truncate and cuspidate; edges often curved downwards; petioles and midribs beneath with rather strong hooked prickles; stipules filiform.

Flowering shoot from fuscous ashy scales. *Prickles* often very strong, deflexed or declining and as well as the shoot white-felted, sometimes also with rather many patent hairs. *Leaves* quinate or ternate, like those of the stem. *Panicle* long, narrow, leafy below, felted, hairy, (very rarely a short seta may be found on the panicle or even calyx); lower axillary branches few, many-flowered, corymbose, short, ascending; upper forming a raceme, patent. *Sepals* white-felted, ovate-attenuate, leaf-pointed. *Petals* pink, obovate, clawed, blunt, jagged. *Filaments* whitish. *Anthers* greenish. *Styles* purple. *Primordial fruitstalk* longer than the calyx. *Fruit* of many small acid illflavoured drupes.

The leaflets of this plant are usually deflexed and rather wavy at their edges; sometimes folded at the midrib so as to be channelled above, but even then usually having ulti-

9—3

mately deflexed edges. The felt of the panicle is white or
more commonly ashcoloured. Sometimes the filaments and
styles are dark red, and the petals deeply coloured.

The *R. fruticosus* of the *Linn. Herb.* consists of bits of
this and of several other species. The present plant is not
the *R. fruticosus* of the Swedish botanists, nor of *Linnæi
Flora Suecica* (ed. 2. 172), where the leaves are described as
being green on their underside. For further remarks upon
the Linnean plant see *R. plicatus.*

French specimens named *R. discolor* are the same as our
plant. Boreau states that the stem is "élevée," but in other
respects his plant and ours agree.

Swedish specimens of *R. discolor* from Fries and Arrhenius have their leaves greyer and more hairy on the under
side, but in other respects closely resemble our plant. Specimens from Denmark, sent by Mr J. Lange, are different:
one from "sepes prope Soro, Sjællandiæ" is probably the *R.
vestitus β viridus* of his *Danske Flora* (ed. 2. 346), but it
does not agree well with our *R. leucostachys:* another from
Jutland is doubtful, but certainly not *R. discolor.*

The *R. discolor* of Wirtgen (*Rub. Rhen.* No. 15) does
not agree with our plant nor with that of the *Rubi Germanici;* but is probably a state of *R. discolor.*

Much confusion exists between *R. discolor* and *R. thyrsoideus,* and preserved specimens are often so similar as to be
nearly undistinguishable, although the plants are truly distinct species. The direction of their stems is totally different: if both plants grow in an open place, the stems of *R.
discolor* will mostly lie quite prostrate after they have formed
a short low arch next to their root; those of *R. thyrsoideus*
ascend highly so as to be suberect during the summer, but
in the autumn grow at the end and descend until they
reach the ground. When the plants grow in hedges or
thickets, or are so strong as to form thickets of themselves,

this difference is not so apparent: then the stems of *R. discolor*, which naturally only rise into a low arch at their base, finding support, continue to raise themselves higher and higher until rather late in the season of growth, when they make a vigorous attempt to reach the ground by sending off one or more slender quick-growing shoots from their extremity. When in this supported state they may easily, but erroneously, be supposed to possess the same tendency to rise as exists in those of *R. thyrsoideus;* but even in these cases the difference in habit is apparent to a careful observer. The stems of *R. discolor* seem to lie along the top of the hedge or bush; those of *R. thyrsoideus* to stand of themselves. The stem of *R. thyrsoideus* is usually much the most sulcate; that of *R. discolor* being often only angular. The panicles present to the eye considerable difference, although it is nearly impossible to describe in what it consists. The edges of the leaflets of this plant have a tendency to turn downwards so as to render the leaflet convex, and often do so in a very marked manner; those of *R. thyrsoideus*, if not flat, turn their edges upwards. The colour and consistence of the felt on the under side of the leaves is very different in the two plants. The petals are different in colour and shape. The styles purple in one are green in the other.

The *R. abruptus* (Lindl. ed. 1) is combined without remark with his *R. fruticosus* in edition 2. In this he is doubtless correct. The specimen named *R. abruptus* by Lindley, from the Hort. Soc. garden and one from the same garden (in the Herb. Borr.), which was called *R. cuneifolius* (Lindl.) in 1829, is clearly the plant named *R. abruptus* by him in the *Synopsis.* It is remarkable that a few years afterwards he should have given the name of *R. rhamnifolius* to specimens of the same plant sent to him by Leighton.

Specimens named *R. Saulii* (Ripart) "sans erreur" by
M. Genevier are I believe identical with *R. abruptus*
(Lindl.), and if separated, the plant ought to bear Lindley's
name; but Mr Baker considers a plant from Wass in Yorkshire to be identical with the *R. Saulii*, and that appears to
me to be a state of *R. leucostachys*.

M. Genevier identified our typical *R. discolor* with his
own *R. Bastardianus* and the *R. rusticanus* of Mercier. I
am unable to see the difference between them, and Mercier's
own specimens are almost exactly the *R. abruptus* Lindl.

I am also unable to separate the *R. cuneifolius*, *R.
elongatus* and *R. undulatus* of Dr Mercier (l. c.) from our
R. discolor.

In the *Phytologist* (iii. 174) Leighton describes four
varieties of *R. discolor*, exclusive of the var. *lividus* (Bab.)
which is now known to be *R. thyrsoideus*, viz.—var. *a*, the
R. fruticosus of Bloxam's *Fasciculus* (No. 9), where the stem
is thinly covered with minute stellate hairs and has declinate or deflexed prickles; and also long spreading but not
very abundant hairs arising from amongst the felt on the
panicle. This is the most common and typical state of
R. discolor.—var. *β*, stem nearly glabrous but with a few
minute stellate hairs and glaucous, prickles nearly patent
and straight, panicle more hairy than in var. *a*. This form
scarcely differs from the preceding.—var. *γ*, stem with a few
scattered weak spreading hairs, deflexed or declinate prickles
and a very hairy panicle. It may be the *R. speciosus*
(Müller) and is the *R. discolor γ macroacanthus* of Bloxam's
Fasciculus No. 11.—var. *δ argenteus*, stem very thickly
covered with minute stellate hairs, prickles declinate,
leaflets rather softly white-felted beneath, panicle very hairy.
This is the *R. discolor γ argenteus* of Leighton's *Shropshire
Rubi.*—The two former of these plants undoubtedly are *R.
discolor*, and scarcely distinguishable: the third seems to

belong with almost equal certainty to *R. thrysoideus:* and the last is, I think, certainly a form of *R. discolor,* but we may be permitted to doubt concerning its identity with the *R. argenteus* (W. and N.).

β *pubescens* (Garke?); caule angulato sellato-sericeo laxè adpressi-piloso, aculeis tenuibus e basi dilatata oblonga depressa vix compressa subito patentibus deflexisve.

R. discolor β *pubescens* Garke Fl. v. Deutschl. ed. 7. 121 ? Metsch l. c. 152 ?

R. pubescens Wirtg.! Rub. Rhenan. No 13.

R. brachyphyllos Müll.! in Wirtg. Rub. Rhenan. No. 128.

Stem angular with flat sides or slightly hollowed on the autumnal shoots, striate, bearing many rather adpressed hairs as well as much stellate down. *Prickles* many, straight or decurved, often small, from an oval rather depressed base. *Leaflets* pilose above, hairy and with dense grey felt beneath; variable in form but usually all cuspidate and narrowed gradually from much above their middle to their base (but sometimes upon the same bush they are oblong-obovate), unequally or even doubly serrate.

Flowering shoot with smaller prickles and rather densely and patently hairy. *Panicle* like that of the typical plant, but usually much longer and therefore relatively narrower, scarcely wider at its base where there are a few axillary branches than near to its top (not pyramidal, as it is called by Dr Metsch); rachis usually having a few often rather many short setæ. Dr Metsch says that the German plant has none.

This plant approaches closely to *R. leucostachys.* It seems to be connected with the typical *R. discolor* by the var. *argenteus* of Leighton. Some remarks will be found under *R. leucostachys.*

The Rev. F. J. A. Hort gathered it at Piercefield, Monmouthshire; the Rev. W. H. Purchas at Penyard near Ross, Herefordshire. I found what seems to be the same plant at Llanwarne, Herefordshire, in company with a bush bearing broader and thinner leaves of which the older and larger are pale green and hairy on the veins but without felt beneath, but those growing upon the secondary stems are densely grey-felted beneath. Mr Purchas also found, at Alton Court Wood near Ross, a bramble which seems to belong to this variety; but it has an enormous quantity of white meal on both of its stems, in addition to the hairs and stellate down. The *R. pubescens* of Wirtgen as represented by his specimen has a nearly or perhaps quite naked stem but agrees very well in other respects with my plant and that of Mr Hort.

R. brachyphyllos (Müll.) is very nearly, if not exactly the same plant.

Habitat. Hedges and thickets. Perhaps our most abundant species. July, August.

Area.—1 2 3 4 5 6 7 8 9 10 11 12 13 . . 16 . . 19 . 21 . 23 24 . . . 28 . 30.

Localities.—i. Ilfracombe, *N. Dev.;* Bath, *N. Som.;* Bonniton near Dunster, *S. Som.* (Coleman).—ii. Poole, *Dors.;* Quarr wood, *I. of W.;* Henfield, *W. Suss.* (Borrer!). —iii. Hook near Thames Ditton, *Surr.;* Horsenton and Notting Hill, *Middl.;* Epping Forest, *S. Essex* (E. Forster!). —iv. Fakenham, *W. Norf.;* Hitcham, *W. Suff.;* very common in *Cambr.; Northamp.* (Bloxam).—v. Llanrumney and Piercefield, *Monm.;* Browberrow, *W. Glouc.;* Shrewsbury, *Salop.; Warw.;* Harlaston (Bloxam), near Stafford, and Rugeley, *Staff.*—vi. Tenby, *Pemb.; Cardigan;* New Radnor, *Radn.*—vii. Llanberis, Pen Maen Mawr, and Capel Curig, *Caern.;* Capel Garmon, *Denb.;* Dolgelly and Pennal, *Merion.*—viii. Twycross, *Leic.;* Stapenhill, *Derby* (Hind!).—

ix. Chester and Bowdon, *Chesh.* (G. E. Hunt!).—x. Thirsk, *N. E. York.*—xi. Newcastle, *Northum.* (Winch !).—xii. Douglas, *I. of Man.*

xiii. *Ayrshire* (Balfour !).—xvi. Lag in Arran, *Clyde Isles.*

xix. Muckross, *S. Kerry.*—xxi. *Kilkenny.*—xxiii. New Grange, *Meath.*—xxiv. Castle Taylor, *E. Galw.* (A. G. More !). —xxviii. *Armagh* (D. Oliver !).—xxx. Belfast, *Antr.* (Hind !).

13. R. thyrsoideus Wimm.

R. *caule erecto-arcuato* angulato sulcato *subglabro,* aculeis è basi valde dilatata compressa declinatis vel deflexis, foliis quinatis, *foliolis* planis subcoriaceis supra glabris *subtus hirtis viridi-cano-tomentosis,* foliolo terminali cordato-ovato vel-subobovato acuminato, paniculæ elongatæ thyrsoideæ ramis inferioribus axillaribus multis paucifloris aculeis validis uncinatis, *calyce tomentoso hirto.*

R. *thyrsoideus* "Wimm. Fl. v. Schles. 204 (1832)." Arrh.! 28. Fries! Summa, 165. Bab.! Man. ed. 3. 95; ed. 6. 109. Blox.! in Kirby, 45. Godr. in Fl. Fr. i. 547; Fl. Lorr. ed. 2. i. 241. Bor. Fl. Centre, ed. 3. 202. Billot! Fl. Gall. et Germ. exsic. No. 1866 (sp.). Wirtg.! Herb. Rub. Nos. 69—73 (sp.).

R. *thyrsoideus a candicans* Bluff et Fingerh. ed. 2. i. pt. 2. 192. Sond. Fl. Hamb. 274. Metsch in Linnæa xxviii. 125. Merc.! in Reut. Cat. Genev. 284.

R. *fruticosus* Rub. Germ. 24. t. 7. Ser. in DC. Prod. ii. 560.

R. *fruticosus* ε *geminatus* Lees! in Steele, 57.

R. *discolor* β *thyrsoideus* Bell Salt.! in Phytol. ii. 104. Bab.! Syn. 14; in A. N. H. xix. 84; Man. ed. 2. 99.

R. *discolor* γ Leight.! in Phytol. iii. 174.

R. *discolor* γ *macroacanthus* Blox.! Fasc. No. 11. (sp.)

R. *candicans* Reichenb. Fl. excurs. 601 (1830). Wirtg. Herb. Rub. No. 5. (sp.)

R. *argenteus* Lees! in Steele, 59; Malv. 56.

R. *vestitus* β *diversifolius* Lees! in Steele, 57.

R. speciosus Müll.! in Flora (1858), 135; in Pollichia, xvi. 93; in Wirtg. Herb. Rub. No. 77 (sp.). Billot, Fl. Gall. et Germ. exsic. No. 3073 (sp.).

R. coarctatus Müll.! in Wirtg. Herb. Rub. No. 120 (sp.).

Stem arching highly or nearly suberect with a descending autumnal shoot from its end, angular, furrowed, with a few hairs; "somewhat hairy and roundish and with short straight conical prickles at the base." *Prickles* strong declining or a little deflexed, from very large compressed bases. *Leaves* quinate, concave as a whole. *Leaflets* nearly flat, wavy and a little turned up at the edges, doubly dentate-serrate, pilose above, greenish-white hairy and softly (but often very finely) felted beneath, not overlapping; basal and intermediate lanceolate; terminal long-stalked, ovate or obovate-acuminate, subcordate at the base; under side of midribs and unfurrowed, petioles with hooked prickles; stipules linear or linear-lanceolate.

Flowering shoot from ashy scales. *Prickles* strong, deflexed. *Hairs* spreading. *Leaves* quinate, like those of the stem. *Panicle* long, narrow, hairy, felted; branches short, rather distant, mostly axillary, patent, corymbose; floral leaves ternate, basal leaflets usually with a large lobe and lobate-serrate externally, or leaves simple and more or less 3-lobed. *Sepals* ovate-acuminate, hairy, felted, reflexed, with a slightly flattened point. *Petals* rather distant, broadly ovate, entire, or finely toothed, blunt, narrowed to the base, white.

Filaments white. *Anthers* faintly fuscous. *Styles* green. *Primordial fruitstalk* as long as the sepals. *Fruit* of rather few subacid drupes. *Seeds* ½-ovate, rather gibbous on the upper part of the inner edge; sides convex.

This species varies considerably and is doubtless often mistaken for *R. discolor*, under which head some remarks upon their differences will be found.

The *R. argenteus* of Mr Lees, from the Cotswold Hills, is not distinguishable from this species; that gathered by him upon Broadheath in Worcestershire is *R. discolor*. The true *R. argenteus* of France and Germany is more nearly allied to *R. discolor* than to *R. thyrsoideus*, but seems to be distinct from both of them.

The *R. macroacanthus* of Mr Bloxam's *Fasciculus of Rubi* is a form of *R. thyrsoideus*. It has hairs upon the upper surface of the leaves but differs very slightly in other respects. The following are the points in which the bush raised from seed sent by Mr Bloxam does not exactly agree with the above description of *R. thyrsoideus*. *Stem* with short, deflexed prickles. *Leaflets* sometimes rather imbricate, hairy above, deeply dentate-cuspidate; terminal roundly oblong, narrowed at both ends, or obovate-acuminate. *Flowering shoot* very hairy. *Leaves* ternate. Lower *leaflets* very unequal and broad on one side of the base; terminal broadly oblong; all hairy above and hairy and felted beneath; floral leaves often simple and lanceolate. *Panicle* short, mostly ultra-axillary, racemose. *Petals* broad, roundish, toothed, pale pink. *Styles* greenish or very slightly pink at the base.

Neither this plant nor the similarly named form of *R. discolor* agrees with the *R. macroacanthos* of the *Rubi Germanici*. A plant gathered by Mr H. C. Watson on the Railway bank at Thames Ditton was named *R. macro-acanthus* by Mr Bloxam and, considering his use of that name, the determination is perhaps correct notwithstanding the fact of its panicle bearing straight or slightly declining prickles. They seem to be the *R. robustus* (Müll.); nevertheless the *R. macroacanthus* of Bloxam's *Fasciculus* is apparently the *R. speciosus* of that German botanist. The leaves of Mr Watson's specimens are often very coarsely dentate, the panicle loose, and all the prickles enormous.

In addition to the *R. robustus* and *R. speciosus*, Müller's *R. coarctatus*, *R. sericophyllus*, and perhaps some of the other plants described in his *Monograph*, are included by us under *R. thyrsoideus*.

Müller informs us that the specimens in Wirtgen's *Rubi* numbered and named, 4 *R. rhamnifolius*, 5 and 33 *R. candicans*, 39 *R. villicaulis*, 53 *R. macroacanthus*, 69, 70, 71, 72, and 73 *R. thyrsoideus*, belong to his *R. speciosus:* they all appear to be forms of our *R. thyrsoideus*, where I also place his *R. macroacanthus v. oblonga* (No. 10). Specimens called *R. fastigiatus* by Mr Wahlberg and gathered near Metz in France are almost exactly our *R. thyrsoideus*.

The varieties of *R. fruticosus* called γ *thyrsoideus*, δ *macroacanthus*, and ε *geminatus* by Lees in *Steele's Handbook* belong to this species.

I have not been able to obtain access to the original *Flora von Schlesien* of Wimmer, and quote it on the authority of later editions. Doubtless the name adopted from some manuscript of Weihe, and used by Reichenbach in his *Fl. excursoria*, is older than that which is now universally employed for this species; but Arrhenius seems to me to have shown good reason for following Wimmer, rather than Reichenbach, in this matter.

Habitat.—Hedges and thickets. July, August.

Area.— . 2 3 4 5 . 7 8 . . . . 30.

Localities.—ii. Selborne, *N. Hants.*—iii. Claygate and Thames Ditton, *Surr.;* Harrow, *Middl.* (Hind!).—iv. *Cambridge;* Sandy, *Beds.;* West Haddon, *Northamp.* (Bloxam). —v. Naunton, *E. Glouc.;* Lydney, *W. Glouc.;* Llanrumney and Red Brook, *Monm.;* Llanwarne, *Heref.;* Alfrick Hallow, Malvern, Broad Heath, and Leigh Linton, *Worc.;* Stoke and Hartshill, *Warw.;* Shrewsbury, *Salop.;* Harlaston, *Staff.* (Bloxam).—vii. Llanberis, *Caern.*—viii. Twycross, *Leic.;* Clifton Campville, *Derby* (Bloxam).

xxx. Near Ben Evenagh, *Derry* (D. Moore!).

b. Sylvatici. Aculei caulis mediocres, sæpissime æquales; pubescentia (densa) piloso-villosa, patens. Folia subtus viridia vel raro albo-tomentosa.

Usually the plants included in this group have neither aciculi nor setæ upon their barren stems, but occasionally a few of each may be found. *R. leucostachys* β *vestitus,* R. *Salteri* β *calvatus* and the form of *R. villicaulis,* which was considered as *R. vulgaris* by Lindley, are sometimes furnished with them in tolerable abundance. The *Sylvatici* and *Spectabiles* do not admit of any satisfactory separation.

14. R. leucostachys Sm.

R. *caule arcuato-prostrato* angulato *piloso-villoso* tomentoso, aculeis multis e basi dilatato-compressa subpatentibus tenuibus, foliis quinatis, *foliolis* planis *subtus* hirtis micantibus *mollibus fulvo-albove-tomentosis*, foliolo terminali obovato ovato rotundatove cuspidato, *paniculæ elongatæ* tomentosæ hirtæ setosæ angustæ ramis brevibus paucifloris aculeis tenuibus declinatis vel angulato-deflexis, sepalis viridi-tomentosis hirtis setosis aciculatis.

a. verus; caule arcuato-prostrato, aculeis plerisque in angulis caulis incertis æqualibus, foliolis coriaceis obovatis sublobato-serratis subtus fulvo-albove-tomentosis hirtis micantibus.

R. leucostachys Sm. Eng. Fl. ii. 403 (1824). Lindl.! Syn. ed. 1. 93. Borr.! in Eng. Bot. Suppl. t. 2631; in Hooker, ed. 2. 246; ed. 3. 249. Bell Salt.! in Phytol. ii. 105. Bab.! Man. ed. 1. 94; ed. 6. 109; Syn. 15. Lees! in Steele 57. Johnst.! East. Bord. 68. Syme's Eng. Bot. iii. 172. t. 448.

R. rudis γ *Reichenbachii* Bell Salt.! in A. N. H. xvi. 368.

R. leucostachys v. argenteus Bell Salt.! in A. N. H. xvi. 366. Bab.! Syn. 15.

R. villicaulis β *argenteus* Bab.! Man. ed. 1. 95.

R. vestitus Lees! Malv. 54. Garke Fl. v. Deutschl. ed. 7. 121.

R. vestitus γ *argenteus* Lees in Steele, 57.
R. argenteus Bab.! Prim. Fl. Sarn. 31.
R. conspicuus Müll.! in Wirtg. Herb. Rub. No. 133. (sp.).

Stem arching slightly at the base, afterwards prostrate, unless supported round below, angular upwards, covered with loose spreading mostly clustered hairs and stellate down; rarely there is an aciculus or seta. *Prickles* many, straight, slender, patent or very slightly declining, a little compressed, from a dilated compressed base. *Leaves* quinate, slightly pedate. *Leaflets* flat, dark green, and slightly pilose above, greyish or yellowish white soft shining hairy and felted beneath, unequally and rather lobate-serrate, sometimes wavy at the edge; lower very shortly stalked, obovate; intermediate and terminal obovate ovate or roundish cuspidate; terminal usually cordate at the base; petioles flat above, and as well as the midribs with rather strong hooked prickles beneath; stipules linear-lanceolate.

Flowering shoot from greenish scales, hairy, with many long straight slender long-based declining prickles, of which those near the base of the shoot are very small; a few short setæ. The prickles are sometimes neither truly deflexed nor declining, but bend downwards at an angle at a little below their middle. *Leaves* ternate. *Leaflets* oblong, pale green beneath; uppermost floral leaves often simple, broad, cordate at the base, three-lobed. *Panicle* long, narrow, hairy, felted, setose, aciculate; axillary branches few, short, few-flowered, corymbose, ascending, distant; ultra-axillary part usually long, often dense, with very short patent corymbose branches; prickles slender, declining, those on the peduncles sometimes deflexed. *Sepals* ovate, acuminate, hairy, felted, with purple setæ and aciculi, reflexed; point long, linear. *Petals* distant, oval, rather acute, toothed, pinkish. *Filaments* nearly white. *Styles* greenish.

Primordial fruit-stalk shorter than the sepals. *Fruit* pur-plish-black. *Nut* broadly half-ovate, truncate below; inner edge nearly straight, slightly rounded at the top.

This plant is generally tolerably well marked, and con-stant when growing in exposed places. But sometimes it wants much of the hair, and has more but looser felt on the stem; also a more open panicle with longer axillary branches. It is then the *R. leucostachys v. argenteus* of Bell Salter and of my *Synopsis*. Occasionally a similar open panicle is accompanied by a thickly clothed stem, which is neither truly hairy nor felted; for the minute stellate hairs that form felt have disappeared, and long clustered exceedingly spreading hairs have taken their place, and almost form a coat of loose felt. This state is found about Malvern by Mr Lees, and seems to be his *R. vestitus γ argenteus*.

The *R. leucostachys* of Dr Johnston's *Eastern Borders* seems to form a connecting link between the typical form and the *var. vestitus*. Its panicle is more like that of the latter, and its stem is more furrowed than is usual even in the most angular forms of the true *R. leucostachys*.

β *vestitus;* caule arcuato, aculeis inæqualibus spar-sis, foliolis cordato-subrotundis cuspidatis irregulariter dentatis subtus pallide viridibus.

R. leucostachys Lindl.! (Hort. Soc. Gard. spec.).

R. leucostachys β *vestitus* Bell Salt.! in Phytol. ii. 105; Bromf. Fl. Vect. 157. Bab.! Syn. 15; Man. ed. 2. 99; A. N. H. Ser. 2. ii. 38. Leight.! in Phytol. iii. 175. Blox.! Fasc. (sp.).

R. vestitus Weihe in Rubi Germ. 81. t. 33 (1825?). Lees! in Steele 57. Blox.! in Kirby 44. Sond. 278. Godr.! Mon. 17; Fl. de Fr. i. 541. Bor. Fl. Centr. ed. 3. 194. Metsch in Linnæa, xxviii. 155. Lange! Danske Fl. 345.

Billot! Fl. Gall. et Germ. exsic. No. 2450 (sp.) Wirtg.! Herb. Rub. No. 84 (sp.).

R. villicaulis Leight.! Fl. Shrop. 231. Bab.! Man. ed. 1. 95 (excl. var. β, γ, δ, and ε).

R. diversifolius Lindl.! ed. 1. 93 (1829).

R. Leightonianus Bab.! in A. N. H. ser. 1. xvii. 240 (1846); Syn. 18; Man. ed. 2. 101. Leight.! in Phytol. iii. 176.

R. sylvaticus β *villicaulis* Lees! in Steele 57.

R. vinetorum Holandre! "Fl. de Moselle ed. 1. 267." (1829).

R. rudis γ *Reichenbachii* Bell Salt.! in Bot. Gaz. ii. 125; in Fl. Vect. 158.

R. conspicuus Müll.! in Flora 1858; Wirtg. Herb. Rub. No. 85 (sp.).

R. macroacanthus Wirtg.! Herb. Rub. No. 9 (sp.).

Stem arching much more than in the typical *R. leucosta-chys*, and usually much rounder, with frequently a few aci-culi and setæ. *Prickles* not wholly confined to the angles of the stem, rather unequal, i.e. although most of them are of equal size, and on the angles, nevertheless here and there a smaller prickle may be found, which is usually (perhaps al-ways) seated on the face. *Leaves* often ternate by the cohe-sion of the lateral leaflets, when quinate they are usually pedate. *Leaflets* rather thin, but coriaceous, broad, obovate or roundish, unequally or doubly dentate, hairy and very finely felted beneath. *Panicle* often with very many sunken purple setæ.

This is the form under which the species is usually found when growing in shade. The original *R. Leightonianus* has a still rounder stem, and broader, thinner, more flexible, and rounder leaflets. That extreme form may be traced, in woods where it abounds, through all the intermediate forms, to the true *R. leucostachys* inhabiting the exposed spots surrounding the wood. The thin leaves of the *R. Leightonia-nus*, and other states of the species, are sometimes nearly naked on the under side; the long hairs being few, and the felt represented by a thin coat of very short recurved hairs. Such plants often much resemble some of the allied species, and are not easily distinguished from them by technical characters. To the practised eye they present less, although still considerable, difficulty.

In rare cases the aciculi and setæ on the barren stem are tolerably abundant, and the plant would, to a casual observer, seem to belong to the *Radulæ* or even the *Glandulosi*. But in every other respect these plants present the true characters of *R. vestitus*. A specimen gathered by Leighton, near Shrewsbury in 1847, is the most marked English example that I have seen. The *R. vestitus* of Wirtgen (*Rub. Rhenan. No.* 16) has this armature well marked. It is very nearly my former *R. Leightonianus*, and has its leaves almost naked beneath. No. 17 of that collection is an extreme example of the same plant as changed by living in much shade.

The *R. diversifolius* (Lind.), as described in the first edition of his *Synopsis*, and formerly cultivated under his eye in the Horticultural Garden (from whence I have seen an authentic specimen), is the *R. vestitus:* the plant bearing the same name in the second edition of his *Synopsis*, and so named for Leighton, will be found described as *R. diversifolius* on a future page, amongst the *Glandulosi Kœhleriani*. This very remarkable change, made unknowingly, in the application of a name has been the cause of not a little difficulty, and even of the use of rather hard words. Each writer naturally believed that the evidence in favour of his own view of the question was uncontrovertible; in one case being founded upon authentic specimens gathered from the bush, to which Mr Borrer was referred for them by Dr

Lindley himself: in the other derived from specimens sent by Mr Leighton to the latter botanist and returned with the name of *R. diversifolius* attached to them by him. The remark in the second edition of Lindley's *Synopis*, in which he rather strongly expresses his astonishment at Mr Borrer's opinion concerning the plant, is quite justified from his point of view, if we bear in mind this singular transfer of the name from one of the *Villicaules* to a plant belonging to the *Glandulosi*; but Mr Borrer's opinion was equally well founded. After the above-mentioned difficulty had been removed I was myself the originator of another: for, having observed an extreme form (as I now consider it) of *R. leucostachys β vestitus* in woods, and being then ignorant of the full effect of shade upon brambles, I thought that it was a distinct species, and called it *R. Leightonianus*. This mistake was the cause of much correspondence and perplexity; but ultimately Mr Leighton himself showed that the plant named in his honour is only the wood-form of *R. leucostachys*. Mr Leighton's remarks will be found in the *Phytologist* (iii. 176), and some of my own in the *Annals N. H.* (Ser. 2. ii. 38).

But we have not yet done with the difficulties which have arisen from forms of this species. In the *Annals N. H.* (xvi. 368) Dr Bell Salter notices a supposed variety of *R. rudis* as the *R. Reichenbachii* of the *Rubi Germanici;* in my *Synopsis* I adopted his views and followed them also in the second and third editions of my *Manual*, referring in the latter of those editions a plant found near Bangor, Caernarvonshire, to that variety of *R. rudis*. A careful examination of tolerably good, but cultivated, specimens of Dr Salter's plant, for which I am indebted to my lamented friend himself, has now convinced me that it also is a form of *R. leucostachys* on the barren stem of which a very few aciculi and setæ show themselves. It is now well known that the presence of a few such arms is not very unusual upon the stems of some species placed in the section *Villicaules*. The plant found near Bangor will perhaps maintain its claims to a place under *R. rudis*.

There is no doubt that the present species is the plant intended by the name of *R. leucostachys* by Smith. It is singular that no plate in the *Rubi Germanici* exactly represents it, and that it only appears there in the wood-form called *R. vestitus*.

Dr Metsch adopts *R. vestitus* as the type of the species, but I am unable to follow his example; first, because I believe that the priority of publication is in favour of Smith's name, and secondly, because the *R. leucostachys* is the more decided form, that called *R. vestitus* being manifestly the effect of shade. He distinguishes *R. vestitus* from *R. pubescens* by its stem being obtusely angular even at the top, not sulcate, and its pubescence patent and dense, and even almost woolly and having sometimes a few setæ interspersed, whilst that of *R. pubescens* consists of adpressed hairs never having setæ intermixed: by the leaflets of *R. vestitus* having a dull green upper side, instead of the lively green of *R. pubescens*: by the panicle being more thyrsoid, densely but loosely felted and setose; that of the latter being almost pyramidal, finely felted and without setæ (in which respect his descriptions do not agree with my plant which has as thyrsoid and as setose a panicle as *R. vestitus*): by the erect straight declining prickles; *R. pubescens* having them usually more or less deflexed. He adds that the prickles are glabrous at the base in *R. vestitus*, which I do not find to be the case in England.

The specimens of *R. leucostachys* sent by Leighton to Nees v. Esenbech were all named *R. villicaulis* by him, and a note appended that in his opinion *R. macroacanthus*, *R. pubescens* and probably *R. sylvaticus* are forms of that

species. An examination of the specimens and comparison
of them with the plates in *Rubi Germanici* convinces me
that they are nearly related to *R. vestitus*. The *R. macroacanthus* of the German authors may perhaps be a form of
R. thyrsoideus but does not much resemble either of the
plants which have been called *R. macroacanthus* in Britain.
R. villicaulis and *R. sylvaticus* are noticed elsewhere.

Mr Sonder remarks that my *R. villicaulis* is very closely
allied to *R. vestitus*, and certainly many of the plants to
which I used to apply that name are really *R. vestitus*.
They have usually thicker leaflets than the plant first so
called in England, and therefore more approach the typical
R. leucostachys. One of them, gathered at Llanberis in
Caenarvonshire, has an enormous compound panicle all the
lower branches of which form secondary panicles similar in
form and size to the primary panicle, or even larger and
more compound than that part usually is.

Dr D. Moore has allowed me to examine a specimen,
gathered by the side of the river Foyle near Londonderry,
which agrees very well with the description and figure of
R. pubescens to be found in the *Rubi Germanici*. The
young part of its stem is exceedingly hairy, but the older
portion is nearly naked. The prickles are large, strong,
more or less deflexed, red with yellow tips, from a long compressed base. The terminal leaflet is obovate-lanceolate
acuminate, not in the least degree cordate and much more
narrowed below than is represented on the plate; it is
acutely and doubly serrate. The panicle is very like that
figured by the German authors but even more leafy. Its
lower part may be wanting in this specimen, but it has
seven very short axillary branches, of which the lower are
racemose and rather distant and the upper corymbose.

The ultra-axillary part of its panicle is very short, as
are also its corymbose branches. But the most conspicuous

difference from the German figure is found in the very decidedly leaf-pointed sepals of the Irish plant. Weihe and Nees describe and figure the sepals as acute, not even possessing the long linear point usually to be found in *R. leucostachys*. I am not inclined to give much weight to this difference, having sometimes had reason to doubt the perfect accuracy of the *Rubi Germanici*, in minute points, and strongly incline to the opinion that Dr Moore's plant is the *R. pubescens* (W. & N.). The identity of that species with our *R. leucostachys* is rendered less certain by the discovery of this Irish plant. Unfortunately only one specimen of it, gathered some years since, has been preserved, and therefore we do not possess sufficient evidence to determine its identity with either of the plants.

In 1848 or 1849 I received from Mr Lees an imperfect specimen of a plant gathered by him on the top of Horsenton Hill in Middlesex, which he supposed to be a form of *R. tomentosus*. It is not safe to attempt the determination of a *Rubus* without a good series of specimens, and therefore I can give no decided opinion upon this plant; but, as far as I am able to judge, it is a state of *R. leucostachys*. It has green felt beneath its leaves, stellate down but very little hair on its stem, more slender conical prickles springing from broader depressed bases than ordinary *R. leucostachys*. The leaves are coarsely and doubly dentate like those of the authentic *R. leucostachys* of Borrer. An accidental aciculus or seta may be found on the stem and the panicle bears an abundance of short inconspicuous setæ.

The specimen of *R. leucostachys* (Lindl.), as obtained from the authentic bush in the Horticultural Society's Garden by Borrer, is *R. leucostachys β vestitus* in its coarser state. The specimens so named by Lindley for Leighton are *R. Lindleianus*.

I possess two specimens of *R. vestitus*, derived from the

Herbarium of the late Dr Leo of Metz, which were named
R. vinetorum by M. Holandre.

Habitat.—Hedges, thickets and woods.   July, August.

Area.—1 2 3 . 5 6 7 8 9 10 11 12 . 14 . . . . 19 .
. . . 24 . 26 . . . 30.

Localities.—i. Monckton Combe, *N. Som.;* Wombwell,
S. Dev. (Briggs!); *E. Corn.* (Hort).—ii. Ryde, Apse Castle,
&c. *Isle of Wight;* Henfield, *W. Suss.*—iii. Long Ditton
and Claygate, *Surr.;* between Panshanger and Bramfield,
Herts.; Halsted, *Essex* (T. Bentall); Harrow Weald Com-
mon and Trent Park, *Middl.* (W. M. Hind).—v. Near
Bristol, and at Coleford, *W. Glouc.;* Llanrumney and near
Monmouth, *Monm.;* Ross, Lanwarne and Much Marcle,
Heref.; Malvern Wells and Little Malvern, *Worc.;* Shrews-
bury, *Salop;* Rugby and Atherstone, *Warw.* (Blox.).—
vi. East Freshwater Bay and Tenby, *Pemb.;* Cardigan;
Radnor; Glan Hafren, *Montgom.*—vii. Bridge of Ogwan
(Borr.!), Llanberis and Menai Bridge, *Caern.;* Capel Garmon,
Denb.—viii. Twycross, *Leic. ;* Chalk Abbey (Bloxam),
Matlock (Borr.!), *Derby.*—ix. Reddish Vale, *S. Lanc.*
(Sidebotham!).—x. Bell Hagg near Sheffield, *S. W. York.:*
Castle Howard (Spruce!), Thirsk, *N. E. York.*—xi. Twizel
House, *Chev.* (G. Johnst.!); Hartley, *Northumb.*—xii. By
the Greta, *Westm.;* Douglas, *Isle of Man.*

   xiv. Penmanshiel, *Berw.* (G. Johnst.).

   xix. Killarney, *S. Kerry.*—xxiv. *E. Galw.*—xxvi. Con-
nemara, *W. Galw.*—xxx. Belfast, *Antr.,* and also in co.
Derry (D. Moore!), Dundonald, co. *Down* (Tate!).
Jersey.

15. **R. Grabowskii** Weihe.

R. *caule arcuato* angulato *subglabro, aculeis multis*
equalibus declinatis deflexisve ad basin valde dilatatam
compressis, foliis quinatis, *foliolis* plicatis supra opacis
glabris *subtus cinereo-tomentosis* irregulariter dentatis
incumbentibus, *foliolo terminali cordato* abrupte cuspidato (ramorum floriferorum ad basin valde dilatato),
paniculæ elongatæ inferne foliosæ ramis ascendentibus
racemoso-corymbosis aculeis multis deflexis, sepalis
cinereo-tomentosis hirtis.

R. Grabowskii Weihe in Wimm. et Grab. Fl. Siles. ii.
32 (1829). Bab.! in A. N. H. xix. 83; Man. ed. 4. 98;
ed. 6. 109. Blox.! in Kirby, 46. Syme Eng. Bot. iii. 173.
t. 449.

R. thyrsoideus γ *apricus.* Wimm. Fl. v. Schles. (1840),
131.

R. carpinifolius Borr.! in Eng. Bot. Suppl. t. 2664
(1830); in Hook. Br. Fl. ed. 2. 244; ed. 3. 247.

R. Wahlbergii β *glabratus* Bell Salt. in Bot. Gaz. ii. 129;
in Hook. Br. Fl. ed. 6. 589 (not Fl. Vect.).

R. nitidus γ *rotundifolius* Blox.! Fascic. (sp.).

Stem "arching," angular or sulcate towards the end,
with a few (often clustered) hairs, often nearly or quite
denuded when old. *Prickles* many, short, rather slender,
often very much declining or towards the end of the stem
deflexed, very much longitudinally dilated and compressed
at their base. *Leaves* quinate. *Leaflets* all stalked, imbricate, plicate, glabrous (or slightly pilose) and dark green
above, hairy and with fine ashy felt beneath, irregularly

dentate (or sometimes the irregularity is very slight except in the direction of the points of the teeth, or the third or fourth tooth is larger than the others); basal oblong-ovate, rather unequal-sided; intermediate nearly round, with a cordate base, abruptly cuspidate; terminal roundly cordate, broader than long, abruptly cuspidate, but at the end of the shoot they are often cordate-oblong-acuminate; midribs and petioles which are scarcely (if at all) furrowed with very many much hooked prickles beneath; stipules very narrow.

Flowering shoot long, nearly glabrous. *Prickles* like those of the stem but often smaller, deflexed or declining, longest and most abundant near to and within the panicle. *Leaves* quinate and like those of the stem, or ternate with the lateral leaflets very broad and with a broad large rounded lobe on the outer side; uppermost floral leaves three-lobed or simple and ovate, cuspidate. *Panicle* long, narrow, hairy but not felted, very slightly setose, very prickly; branches ascending; about three of the lowest axillary, rather long but falling short of the leaves, racemose-corymbose, rather distant; others corymbose and forming a close leafless raceme; peduncles and sepals hairy, felted, aciculate, with many very short pale setæ. *Sepals* oblong, acuminate with a short flattened point, greenish with a narrow white edge externally, loosely reflexed from the oblong black fruit. *Fruit* rarely produced. *Seeds* very broadly $\frac{1}{2}$-ovate; inner edge gibbous below, otherwise straight; sides convex.

I have not seen living specimens of this plant, nor even dried ones of the flowers, nor does Mr Bloxam give us any information concerning the petals, stamens or styles.

The *R. carpinifolius* of *Eng. Bot. Suppl.* (tab. 2664) is probably the same specifically as my *R. Grabowskii*, although it differs in some respects. Its stem is more commonly furrowed and is rather thickly clothed with clustered spreading hairs; its leaflets are proportionally longer, more acuminate, sometimes rather obovate and those of the flowering shoot narowed slightly to their base; more deeply toothed or even jagged. Their panicles are exceedingly alike in nearly all respects; but the leaflets are narrower, and the top of the rachis is felted like the peduncles, in Borrer's plant. Its petals are "pink and the filaments of its stamens dark purple." The fruit is only sparingly produced. Mr Borrer's plant is certainly not the *R. carpinifolius* of the *Rubi Germanici*, and he would "not have ventured to give it as the *R. carpinifolius* (W. & N.), but for the exact accordance of an authentic specimen." It is not stated by whom that specimen was named, and I certainly believe that it had no real claim to the name which it bore. Any person who will compare *tab.* 13 of the *Rubi Germanici* with the plate in *English Botany* must be struck with their exceedingly great difference.

There is another plant which I think may safely be combined with these. It was sent to me by Mr Bloxam, and is growing in the Cambridge Botanical Garden from seeds given by him. He supposed it and *R. Colemanni* to be the *R. infestus* (Weihe). It approaches nearly to the *R. carpinifolius* of Borrer, but its leaves and even leaflets are incurved at the edges, whilst Borrer states that those of his plant have their edges "often somewhat deflexed." Its leaflets resemble those of Borrer's *R. carpinifolius* in not overlapping, in which respect both these plants differ remarkably from the *R. Grabowskii*. It much resembles *R. Colemanni*, but there the leaflets are concave, hairy on the veins but not felted beneath, the petals are oval entire and white, not ovate jagged and pale pink as is their state in Borrer's *R. carpinifolius* and the present plant; neither are its filaments and styles at all pink. This plant grows in Hartshill wood, Warwickshire.

My original *R. Grabowskii* agrees very nearly with the elaborate and excellent description given in the *Flora Silesiæ*, and I therefore quote that work with much confidence. There are slight differences between it and the English plant of which the following are the chief. The stem is called glabrous, and such is often its condition with us when become old. Our plant has long hairs upon the under side of the veins of its leaves, that of Silesia is said to want them. The panicle is described as "ampla, ... pyramidata, apice acuta, usque fere ad apicem foliosa," but it is not so in our plant. Also the expression "glandulæ nullæ" occurs and is probably intended to apply to the panicle: if so there is a very marked difference between the plants; for with us the peduncles and sepals bear an abundance of short (and therefore inconspicuous) setæ or shortly stalked glands.

There can be very little doubt concerning the true place of this plant. Wimmer and Grabowski considered it as much like and probably often called *R. fruticosus* (our *R. thyrsoideus*), and its similarity to some states of that species is considerable; but its habit is said to differ, and there are many points of nonconformity. Wimmer makes it a variety of *R. thyrsoideus* in his *Fl. v. Schles. Prüs. und Oster.* Until we know more about it we shall probably act most wisely if we retain it as a distinct species. Should it prove to differ from the true *R. Grabowskii* (a name which seems to be now unnoticed by German botanists) it might well be called *R. Borreri;* the plant so named by Dr Bell Salter being well known to be only *R. Sprengelii.*

Habitat.—Hedges and thickets. July, August.

Area.— . 2 . . . . 7 8.

Localities.—ii. Near Henfield, *W. Suss.*—vii. Near Beaumaris, *Angl.* (W. Wilson! in Linn. Herb. Brit.).—viii. Near Cadeby, *Leic.;* Hartshill wood, *Warw.*

16. R. Colemanni Blox.

R. caule arcuato angulato subglabro, aculeis multis subæqualibus declinatis ad basin valde dilatatam compressis, foliis quinatis, *foliolis convexis* supra opacis *subtus viridibus in venis hirtis* irregulariter dentatis incumbentibus, foliolo terminali rotundato-cordato-acuminato (ramorum floriferorum rotundato vel late ovali), paniculæ elongatæ inferne foliosæ ramis ascendentibus corymbosis vel ramis axillaribus racemosis aculeis multis tenuibus deflexis declinatisve setis aciculisque multis, sepalis cinereo-tomentosis hirtis.

R. Colemanni Blox.! in Kirby, 38 (1850). Bab.! Man. ed. 6. 109.

R. fusco-ater β Colemanni Bab.! Man. ed. 3. 101; ed. 4. 104.

R. infestus Blox.! MS. (in part, not of Weihe).

Stem arching, angular throughout, with a few aciculi and setæ and scattered hairs; the aciculi often having deciduous glandular tips. *Prickles* on the angles of the stem, nearly equal, strong, but rather slender, short, declining, from a much compressed and long base. *Leaves* quinate or rarely ternate. *Leaflets* broad, all stalked, imbricate, convex, plicate, slightly pilose above, scarcely paler not felted but densely hairy on the chief veins and pilose on the others beneath, irregularly and doubly dentate; basal oblong, acuminate; intermediate obovate, cuspidate; terminal roundly cordate-acuminate; midribs and petioles, which are flat above, with many strong slender slightly curved prickles beneath; stipules linear-lanceolate.

Flowering shoot from fuscous scales, nearly glabrous and with many small setæ and short thick-based aciculi throughout. *Prickles* (as also on the rachis and peduncles) many, unequal, slender, declining or slightly deflexed. *Leaves* mostly ternate. *Leaflets* broad, pale green, hairy and sometimes very slightly felted beneath, pilose above; basal rather unequal-based; terminal nearly round, cuspidate. *Panicle* long; branches few, short, few-flowered, corymbose; many lower axillary, short, racemose; floral leaves often simple, ovate or cordate ovate. *Sepals* ovate, narrowed to a linear point, externally hairy felted setose and aciculate, reflexed. *Petals* distant, oval, clawed, blunt, denticulate, white or pinkish. *Filaments* white. *Anthers* yellowish. *Styles* pale green. *Primordial fruitstalk* longer than the calyx. *Seeds* very broadly half-ovate; inner edge nearly straight throughout; sides convex.

Sometimes the terminal leaflets of the stem are almost exactly cordate; also occasionally some of the lower branches of the panicle exceed the leaves and are very prickly.

My specimen from Coventry is densely clothed with silky hair on the ribs beneath the leaves, and the terminal leaflets on the flowering shoot are oblong: the plant from Packington has much, but very short and adpressed, hair on those ribs, and the same leaflets are nearly round. Neither plant has any felt on its leaves or on its panicle, but the latter part and the peduncles are very hairy.

Mr Bloxam believed that the Coventry plant is the true *R. infestus* (Weihe), in which opinion I cannot agree. It has not the formidable armature of that species as represented in the *Rubi Germanici*. It is true that there are a few gland-tipped aciculi and setæ on this plant (as seen in cultivation at Cambridge from seeds sent by Mr Bloxam), but the stem is nevertheless exceedingly unlike that depicted on tab. 30 of the *Rubi Germanici*. Nearly if not quite as many may sometimes be found on several of the species included in the next Subsection (*Spectabiles*), but this plant does not seem to be closely allied to them in other respects. The panicle indeed is so very glandular and aciculate that it might well pass for that of one of the *Glandulosi*, but such is also the case with some other species not truly ranging with the *Glandulosi*.

Mr Bloxam considered this plant to be closely allied to the *R. Grabowskii*, and their similarity in appearance is certainly very great. It will be seen from the descriptions that the typical *R. Grabowskii* has felted leaves, a much shorter and more abrupt terminal leaflet, and a panicle with exceedingly few setæ and probably no aciculi. Unfortunately there are many points relating to the typical *R. Grabowskii* (from Cadeby) with which we are not acquainted, especially relative to the flower. It is quite possible that these two plants may constitute only one species, but in our present uncertainty it is best to give a full character and description of *R. Colemanni*.

The *R. trichocarpus* (Timer. MS.) of Billot's *Flora Gall. et Germ. exsic.* No. 3076, is perhaps also a form of *R. Grabowskii*. Its leaves are quite the same as those of my typical plant, but there are setæ on the general and partial petioles. The stem has many short setæ and a few aciculi. Its panicle is more setose than that of *R. Colemanni*. The name is derived from the hairy fruit : but that is a very inconstant character. Hairy fruit has been noticed in our *R. Grabowskii*.

It is much to be desired that we knew a little more about these three plants, which probably form one species. They tend to show, as has been already stated, that the *Sylvatici* cannot always be separated from the *Spectabiles*, nor the latter from the *Radulæ*, however much the marked species included in those sections may differ.

There is a plant in Mr Baker's Herb., collected by the late Mr Hailstone at Tarbet by Loch Lomond, which is

named *R. septorum* (Müll.), *R. pubescens* (Bor. not of
Weihe). It resembles *R. Colemanni* in all respects except
that the leaflets of the flowering shoot are lanceolate, and
the panicle is much narrower from all its branches being
very short. If the stem-leaves were felted beneath it might
well be referred to *R. leucostachys* (Sm.). It is well deserving of study by those botanists who may be so fortunate
as to meet with it. I also suspect that the plant named
R. Boreanus by Genevier will best be placed here. All the
panicles of it that I have seen are nearly simple. Its
author thinks it is near *R. leucostachys*, and Baker named it
R. macrophyllus.

Habitat. Hedges. July, August.

Area. . . . . 5 . . 8 . 10.

Localities. v. Near the Railway Station at Coventry,
Warw. —viii. Packington, *Leic.* —x. Wass, *N. E. York.*
(*R. Boreanus*).

17. R. Salteri Bab.

R. *caule* arcuato-prostrato *angulato sulcato sub-glabro, aculeis* e basi dilatato-compressa *subpatentibus tenuibus* compressis, foliis quinatis, *foliolis* tenuibus *grosse et duplicato patenti-dentatis* utrinque viridibus subtus in venis tantum pilosis, foliolo terminali late obovato cuspidato-acuminato basi subcordato, paniculæ longæ laxæ hirtæ ramis ultra-axillaribus brevibus pauci-floris corymbosis patentibus, rachi undulato, aculeis tenuibus declinatis, sepalis hirtis tomentosis erecto-patentibus.

R. Salteri Bab.! Man. ed. 4. 100; ed. 6. 110. Syme Eng. Bot. iii. 174.

a. *Salteri;* foliolis lobato-duplicato-serratis, pani-culæ cylindricæ rachi subrecto ramis corymbosis paten-tibus, sepalis erecto-patentibus.

R. Salteri Bab.! in A. N. H. xvii. 172 (1846); Syn. 10; Man. ed. 2. 97; ed. 3. 93. Bell Salt.! in Bot. Gaz. ii. 119 (excl. var. β); in Bromf. Fl. Vect. 156.

R. acuminatus Genev. in Mem. Soc. Acad. Angers. viii. (1860).

R. fallax Chabois. in Müll. Mon. 82. (1863); Etude du Rub. 20.

"Creeping extensively." *Stem* long, arcuate-prostrate, angular, striate, furrowed, green, with short scattered patent hairs. *Prickles* few, longish, from a long compressed base, nearly equal, declining, confined to the angles of the stem. *Leaves* quinate-pedate. *Leaflets* strongly and doubly dentate-serrate in their upper half, the serratures simple below and decreasing in size downwards, green on both sides, dull and slightly pilose above, rather soft from the many short hairs on the veins beneath; basal obovate-oblong, acute, shortly stalked; intermediate obovate, rather wedgeshaped below, cuspidate; terminal broadly oblong, subcordate below, rather cuspidate; petioles and midribs with a few small strong declining or deflexed prickles beneath; stipules linear-lanceolate.

Flowering shoot long, hairy, from dark brown scales clothed with silky ashy hairs. *Prickles* few, small, deflexed, from very long compressed bases. *Leaves* ternate. *Leaflets* nearly equal, green on both sides, pilose above, more thickly pilose on the veins beneath, obovate-oblong; terminal usually much narrowed below or even wedgeshaped. *Panicle* narrow, compound, hairy, with a few sunken setæ; prickles few, short, slightly deflexed; few lower branches axillary from ternate or three-lobed or simply ovate leaves, often long and patent; other branches short, patent, simple or 2-3-flowered. *Sepals* woolly, ovate, with a long leaflike point, embracing the oblong black fruit. *Petals* lanceolate, narrowed below, white. *Terminal flower and fruit* sessile. *Nut* nearly ½-ovate; inner edge straight.

The plant noticed in my *Synopsis*, from Cramond Bridge, is *R. latifolius*. I have not seen Mr Sidebotham's plant from Bradbury Wood in Cheshire.

M. Genevier names a plant in Baker's Herb. gathered in South Devon *R. calvatus* (Blox.), and decides that it is his *R. acuminatus* and *R. fallax* (Chaboisseau, *Etude du Rubus*, 20). It exactly resembles the original *R. Salteri* from Apse Castle Wood, and therefore he confirms my idea that *R. Salteri* and *R. calvatus* are the extremes of one species. I have not seen the typical *R. Salteri* (Bab.) from any place in Britain except Apse Castle Wood.

β *calvatus;* foliolis grosse dentatis, dentibus distantibus apice recurvatis interstitiis denticulatis, paniculæ rachi flexuoso ramis inferioribus subracemosis et ascendentibus summis corymbosis et patentibus, sepalis laxe reflexis.

R. calvatus Blox.! in Kirby 42 (1850). Bab.! Man. ed. 3. 97; in A. N. H. Ser. 2. ix. 127. Bor. Fl. cent. ed. 3. ii. 199. Genev.! Essai ii. 19.

R. Salteri β *calvatus* Bab.! Man. ed. 5. 101; ed. 6. 110.

Stem slightly arched at the base but chiefly prostrate, angular throughout, slightly furrowed towards the end, with few scattered hairs, rarely a few aciculi and setæ, bright shining red when much exposed; subsessile glands rather plentiful. *Prickles* nearly but not quite confined to the angles of the stem, towards the base of which they are much scattered, unequal, many, slender, slightly compressed, subpatent, from a rather long compressed base. *Leaves* quinate. *Leaflets* stalked, convex, thin, green on both sides, with large recurved and small intermediate teeth, rugose and glabrous above, rough, hard and slightly hairy on the veins beneath; basal with rather long stalks, oblong, acute; intermediate oblong-obovate, subcordate below, rather cuspidate; terminal roundly oblong or slightly obovate, cordate below, subcuspidate; petioles a little furrowed above and together with the midribs having many strong deflexed prickles beneath; stipules linear-lanceolate.

Flowering shoot rather angular, hairy, wavy; prickles many, rather long, slender, longest near to the panicle, declining or slightly deflexed, from a long compressed base. *Leaves* quinate or ternate. *Leaflets* green on both sides, slightly pilose above, more so beneath, finely doubly serrate with the large teeth having patent tips; basal oval, acute. rather unequal-sided; intermediate obovate, rather cordate

12

below, subcuspidate; or basal and intermediate of each side combined into a single roundly oval slightly unequal-sided cuspidate leaflet, which is nearly as large as the terminal leaflet; terminal roundly oval, subcordate scarcely narrowed below, cuspidate. *Panicle* long, often leafy to the top, lax; rachis wavy, and as well as the branches and peduncles felted, hairy, with many sunken setæ and subsessile glands; branches mostly axillary, short, ascending, racemose, uppermost corymbose and patent. *Sepals* oblong, slightly prickly, felted, hairy, a little setose, often leaf-pointed, loosely reflexed from the fruit. *Petals* rather distant, ovate, clawed, pinkish (deep rose-coloured on Irish specimens). *Filaments* pink. *Anthers* yellowish. *Styles* greenish. *Terminal flower and fruit* shortly stalked. *Nut* half-oblong; inner edge straight.

Mr Bloxam long supposed this to be the true *R. sylvaticus* (W. & N.), but that plant seems to be a state of *R. villicaulis.* He afterwards gave a new name to it derived from its stem soon becoming bald. It does not seem to be at all nearly related to *R. villicaulis.*

It is probable that *R. Salteri* and *R. calvatus* form the extremes of one species, although the characters given above might be considered as sufficient to separate them specifically. The serratures of the leaves differ considerably in well-developed specimens, but the general look presented by them is very similar. In *R. calvatus* they resemble large spreading teeth set at equal distances along the edge of a minutely dentate leaflet; in *R. Salteri* they are large unequal-sided teeth which are themselves dentate. It is very difficult, and probably undesirable, to attempt a distinction between these structures. Even in *R. Salteri* the points of the large teeth are sometimes bent backwards, but not in so great a degree as are those of *R. calvatus.* The panicle of *R. calvatus* is sometimes leafy almost to its

top, all its branches being axillary and ascending; that of
R. Salteri is usually leafless in its upper half.

Mr Bloxam is known to deny the propriety of the
combination of R. Salteri and R. calvatus. He thinks that
they are quite distinct species. Until I saw a remark by
Mr Syme (Eng. Bot. iii. 175) this opinion was unaccountable.
Mr Syme states that it results from Dr Salter having given
a specimen of R. Balfourianus to Mr Bloxam as R. Salteri.
It is curious that Dr Salter should have made this mistake,
and thus proved himself to have so slight an acquaintance
with a plant which was first noticed as distinct by himself.
But in the Botanical Gazette he joins R. Balfourianus to R.
Salteri, and continued of that opinion when revising his
arrangement of the species for the British Flora (ed. 7. 125).
It is clear therefore that Mr Bloxam's opinion, otherwise of
the highest value, is in this instance founded on a mistake
made by Dr Salter.

A specimen named R. affinis, forma II, by M. Questier
belongs almost certainly to this species. He compares it
with specimens of R. Salteri, and seems to be much inclined
to consider them as identical. It does not exactly agree
with either of my varieties, but might perhaps be placed
between them. Its panicle seems to remove it widely from
R. affinis.

Mr Lange sent to me a specimen of a plant "in silvis
Fioniæ frequens," which was named R. discolor by Arrhenius. He suspected it to be the R. sylvaticus of the earlier
editions of my Manual. Certainly there was a time when
I should have included Mr Lange's plant under that name.
It is almost exactly the R. calvatus of Bloxam. It will
have been seen under R. discolor that this specimen was
erroneously named by Arrhenius, and that his real R. discolor
and that of Fries is very closely allied to our plant which
bears the same name.

Habitat.—Open woods and hedges.   July, August.

Area.—1 2 . . 5 . 7 8 . 10 . . . . 19 . . . .
. 30

Localities.—i. Bank of R. Erme, *S. Dev.* (Briggs!).—ii. Apse Castle wood, *Isle of Wight.*—v. Almond Park, *Salop;* Sydney, *W. Glouc.*—vii. Tan y Bwlch, *Merion.* (Borr.!).—vii. Twycross, Ashby de la Zouch, and between Loughborough and Wymeswold, *Leic.*—x. Between Thirsk and Dalton, *N. E. York.*—xii. Douglas, *Isle of Man.*

xix. Muckross near Killarney, *S. Kerry.*—xxx. Frequent in county of Derry (D. Moore).

18. R. carpinifolius W. and N.

R. caule erecto-arcuato angulato striato patenti-(fasciculato-)piloso, aculeis e base dilatato-compressa *declinatis tenuibus* conico-compressis foliis quinatis, *foliolis* tenuibus irregulariter sed *argute serratis* pilosis subtus pallide viridibus vel canescentibus hirto-velutinis vel raro viridi-albo-tomentosis, foliolo terminali obovato-acuminato vel cuspidato, *paniculæ angustæ* racemosæ hirtæ setosæ ramis inferioribus axillaribus paucifloris brevibus aculeis deflexis vel declinatis, sepalis hirtis setosis a fructu laxe reflexis.

R. carpinifolius Rubi Germ. 36. t. 13 (1824?). Bab.! Syn. 19; Man ed. 6. 110. Reichenb.! Fl. excurs. 602; Fl. exsic. No. 874 (sp.). Bell Salt.! in Phytol. ii. 107. Syme's Eng. Bot. iii. 175.

R. carpinifolius α Bab.! Man. ed. 4. 99.

R. rhamnifolius (first form in part) Leight.! Fl. Shrop. 226.

R. vulgaris δ *carpinifolius* Metsch in Linnæa, xxviii. 145.

R. rhamnifolius Johnst.! E. Bord. 65.

R. vulgaris Rubi Germ. 38. t. 14.

Stem forming a very large arch so as often to seem suberect in summer, angular, striate, with spreading often clustered hairs and few or no subsessile glands. *Prickles* on the angles, rather slender, compressed, declining, from a very long compressed base. *Leaves* quinate. *Leaflets* flat (?), obovate-elliptic-acuminate or obovate-cuspidate, usually small, finely but irregularly and somewhat doubly serrate, with the teeth very acute and remarkably directed forwards, dull green and distantly pilose above, scarcely paler and often densely hairy on the veins beneath and rarely felted; basal very shortly stalked; terminal rounded or subcordate at the base; petioles, which are probably flat above, and midribs with slender hooked prickles beneath; stipules linear-lanceolate.

Flowering shoot from brown scales clothed with silvery hairs, hairy. *Prickles* few, small, slender, deflexed, from a large compressed base, longest and strongest at about the base of the panicle. *Leaves* ternate or quinate, like those of the stem in all respects; floral leaves simple or three-lobed. *Panicle* compound, racemose, often nearly cylindrical, frequently simple; branches short, few-flowered, corymbose; lower axillary, racemose or corymbose, ascending; rachis and peduncles hairy, felted, with a few sunken setæ. *Sepals* lanceolate, gradually narrowed into a long slender point, patent or slightly reflexed with the flower, reflexed from the fruit, felted, hairy, aciculate. *Petals* ovate-lanceolate, much narrowed below, white or reddish. "*Filaments* pink. *Styles* green." (Baker's MS.). *Primordial fruitstalk* as long as the calyx.

The stem is angular with the sides often quite flat and striate, but in rare cases it is furrowed towards the tip: the hairs are often nearly all scattered and patent; but sometimes they become more in number, form clusters and diverge from their common base; or are even so much reduced in length as almost to pass into felt; or are so few in number as to escape notice unless very carefully looked for. The fine and very acute serration of the leaves usually becomes double towards the tip, where it is sometimes even slightly lobate: the under side is usually quite without felt, having only a dense covering of long hairs which are all seated upon the veins; or a very fine coat of felt occupies the space between the veins; the long hairs always give a very soft feel to the underside.

Specimens gathered in the Lake District by Mr Hort have much fewer and shorter hairs, but a dense coat of whitish felt beneath their leaves. They appear almost certainly to be the *R. vulgaris* (W. & N.), which is probably rightly combined with *R. carpinifolius* by Dr Metsch. I cannot agree with that botanist in joining *R. villicaulis* to them. Dr Johnston's *R. rhamnifolius* agrees with these plants from Mr Hort.

In the most typical specimens the panicle is rather short, nearly simple, cylindrical and leafless, only a few of the lowest branches being conspicuously separated from the rest and from each other; those branches are axillary and fall considerably short of the leaves; they are also racemose, whilst the very short branches of the rest of the panicle are corymbose or 1-flowered. Strong panicles may sometimes be found where the axillary branches are increased in number, at the expense of those which are ultra-axillary, and the whole panicle becomes rather pyramidal. When the panicle is weak the branches are often nearly all reduced to 1-flowered peduncles and the inflorescence is nearly a simple raceme.

I believe that this is the plant named *R. carpinifolius* in the *Rubi Germanici*, although the terminal leaflet is not so uniformly cordate at the base as the authors of that work supposed, neither is it constantly acuminate as they seem to have thought. There is another difficulty attending the identification of our plant with that similarly named by continental authors. Our plant arches so highly that its stem may often be supposed to be suberect : continental botanists describe the stem as arcuate-decumbent. My specimen contained in Reichenbach's *Flora exsiccata* was gathered by Dr Weihe, but is unfortunately very imperfect.

The *R. carpinifolius* of Bell Salter's paper upon the *Rubi* of Selborne (*Phytol.* ii. 107) is clearly this species; but a specimen similarly named by him, and given to me as from that place, seems to be *R. thyrsoideus.*

I have received *R. carpinifolius* from M. Questier with the name of *R. rhamnifolius.*

Although the *R. rhamnifolius* of Borrer is almost certainly the plant already described under that name, which is also the *first form* of the species according to Leighton's views as expressed in his *Flora*, nevertheless the specimens named *R. rhamnifolius* by Borrer for Leighton appear to be *R. carpinifolius.* They have the very hairy stems mentioned by Leighton (*Fl. Shrop.* 227). The *R. carpinifolius* of Leighton will be noticed under *R. Koehleri.*

Habitat.—Open places, preferring hilly districts. July, August.

Area.— . 2 3 4 5 6 7 8 9 10 11 12 . . . 16 . . 19.

Localities.—ii. Poole, *Dors.* (Bell Salter!); *Isle of Wight* (D°.!); St Leonard's Forest, *W. Suss.* (Borrer!).—iii. Long Ditton, Chertsey, *Surr.*—iv. Lowestoft, *E. Suff.*—v. Shrewsbury, *Salop.*—vi. Pont Erwyd, *Card.;* Glan Hafren, *Montgom.*—vii. Llanberis and Penmaen Mawr, *Caern.;* Capel Garmon, *Denb.;* Cwm Bychan, *Merion.*—viii. Twycross, *Leic.*—ix. Bowdon, *Chesh.* (T. Coward!); Accrington, *S. Lanc.*—x. Hebden Bridge, *S. W. York.*—xi. Ford, *Chev.* (Johnston!); Bardon Mills, S. Tynedale, *Northumb.*—xii. Keswick, *Cumb.;* Stock Gill, *Westm.;* Douglas, *Isle of Man.*

xvi. *Bute* (Balfour!).

xix. Turk Mountain near Killarney, and Derrinane, *S. Kerry.*

19. R. villicaulis W. and N.

R. *caule arcuato* vel arcuato-prostrato angulato patenti-piloso, aculeis e basi dilatato-compressa subpatentibus validis conico-compressis, foliis quinatis, *foliolis tenuibus* dentato-serratis subtus pallide viridibus micantibus villosis vel in venis tantum hirtis, foliolo terminali obovato vel cordato-obovato-orbiculato cuspidato subacuminatove, *paniculæ apertæ compositæ* foliosæ hirtæ tomentosæ subsetosæ ramis corymbosis aculeis tenuibus declinatis vel deflexis, sepalis hirtis setosis aciculatis a fructu laxe reflexis.

R. villicaulis Rubi Germ. 43. t. 17 (1825?). Lees Bot. Malv. 57. Bab.! Man. ed. 3. 97; ed. 6. 110. Bor. Fl. Centre, 199. Garke Fl. v. Deutschl. ed. 7. 119 (in part).

R. sylvaticus Bab.! Syn. 16 (in part). Blox.! in Kirby, 43.

R. pampinosus Lees! Bot. Malv. 55 (1852); Bot. Worc. 44. Bab.! Man. ed. 4. 100.

R. vulgaris η *villicaulis* Metsch in Linnæa, xxxviii. 145.

R. vulgaris α *umbrosus* Lange! Danske Flora, ed. 2. 344 (in part).

R. infestus Billot! Fl. Gall. et Germ. exsic. No. 2453 (sp.).

Stem slightly arching, but often with a long prostrate extremity, nearly round at the base, angular upwards; hairs dense, patent, mostly solitary, often very many; rarely a few setæ. *Prickles* subpatent, rather strong, conical, compressed, from a rather long and rather broad base. *Leaves* quinate, nearly flat. *Leaflets* dull green and pilose above, paler green, and with long soft silky hairs on the veins but not felted beneath, slightly convex, irregularly dentate-serrate, often with larger patent teeth at intervals; basal leaflets shortly stalked, elliptic; intermediate obovate, acuminate, bluntly wedgeshaped or rarely subcordate at the base; terminal obovate or cordate-obovate or nearly roundly cordate, cuspidate; petioles flat above, and as well as the midribs having rather strong, nearly straight, declining prickles beneath; stipules linear.

Flowering shoot from brown scales clothed with ashy down, hairy, with rarely an aciculus or seta. *Leaves* ternate. *Leaflets* broad, like those of the stem but more hairy above. *Panicle* long, loose, compound; axillary branches many, usually short, erect-patent, mostly corymbose, a few of the lowest racemose and rarely lengthened into secondary panicles like the primary one; upper part with more patent few-flowered corymbose branches; rachis and peduncles hairy, slightly felted, with short setæ. *Sepals* ovate-acuminate, spreading, hairy, felted, setose, aciculate, with a long rather leaflike point. *Petals* pink, obovate, clawed. *Filaments* pink. *Anthers* greenish. *Styles* greenish with a pink base. *Primordial fruit-stalk* shorter than the sepals.

It is stated in the *Rubi Germanici* that the under side of the leaves is white, and the plant is so figured; but I have not seen any English specimens having such leaves. In all our plants the spaces between the veins on the underside are quite naked and rather pale green; but often so much overhung by the dense long and shining hairs, which clothe the veins, as to be nearly hidden. Weihe and Nees describe the branches of the panicle as all divaricate and corymbose, but figure them as ascending: neither of these states is constant with us; we have specimens like the figure in *Rubi Germanici*, and also some where the lower branches are rather racemose and ascending, whilst those which are ultra-axillary are divaricate and corymbose. Boreau informs us that the petals are white, and they are so drawn in the *Rubi Germanici;* but all the plants that I have seen have pink petals. The plant sent by M. Questier agrees with ours.

Now that the thick and felted-leaved plants are separated from this species and referred to *R. leucostachys β vestitus,* it is tolerably constant in form. It still includes nearly all the plants placed under *R. sylvaticus* in my *Synopsis,* and agrees well with the characters there given.

A plant, gathered at Cowleigh Park near Malvern (which has very round, but still slightly obovate, leaves with a cordate base and a strong cusp, also a loose panicle with long axillary lower branches), was named *R. affinis* by Mr Lees many years since. It seems to be a form of *R. villicaulis,* for the stem, although now nearly naked, shows manifest signs of having once been pilose, and there are also traces of a very few slender aciculi or the bases of the strongish setæ. From this condition of the stem it clearly cannot be *R. affinis.*

I am indebted to Mr Lees for a specimen of what seems to have been an exceedingly large decumbent (or rather probably arcuate-procumbent) plant gathered in Birchin Grove near Worcester. It has very long strong prickles which are rather unequal in size. Its leaves are all ternate with enormous cuspidate leaflets; the lateral shortly stalked, very broad, lobed on the lower edge; the terminal roundly obovate; all coarsely and irregularly dentate, pilose above, hairy on even the finest veins beneath. This I consider to be the form assumed by *R. villicaulis* when growing in deep shade.

A specimen gathered by Mr H. C. Watson near Barwell Court, Surrey, in 1854, has only a distant resemblance to *R. villicaulis,* although apparently a real form of that species.

Its stem is not very hairy and has small scattered unequal prickles, the smaller of which may have been gland-tipped, but if so they have now nearly all lost their heads. Its leaves are nearly naked, small, and finely toothed, but have that approach to a double toothing which is usual in this species. It is an elegant plant, if I may judge from the only specimen that I have seen.

The Rev. W. H. Coleman gave to me a very large form of this species; apparently a wood plant. Here the prickles are much smaller than is usual in the species, the leaves are quinate with leaflets scarcely abnormal except by their very large size. The panicle is enormously long; the joints of the rachis, the branches and peduncles being all much lengthened, the peduncle of the terminal flower alone excepted. It was gathered at Mardly Heath, Hertfordshire.

The *R. pampinosus* (Lees) is not to be separated from this species; nor does it seem distinguishable as a variety. The cordate-ovate-orbicular cuspidate leaflets and usually very large panicle are its chief peculiarities, even in its describer's opinion, and they are found in every state of change until the usual *R. villicaulis* is reached. I was led to suppose that it might be a distinct species from having confounded plants bearing felted leaves with the true *R. villicaulis*. I have received the same form (*R. pampinosus* Lees) from Mr Lange, who gathered it near Frederica in Jutland, and gave it the name of *R. umbrosus* (Weihe). Arrhenius has written upon the ticket "mihi ignotus." But the *R. vulgaris* of Lange's *Flora*, to which he joins the Jutland plant, seems to be my *R. macrophyllus a umbrosus;* for he quotes the specimens called *R. Sprengelii* in Fries's *Herb. Normale* as belonging to the same species, and they appear to be certainly my *R. macrophyllus a umbrosus;* not *R. villicaulis.* M. Questier published the same plant (in Billott's *Fl. Gall. et Germ. exsic.* No. 2453) under the name of *R. infestus* (W. & N.). It seems to my eye to have exceedingly little in common with the *R. infestus*, and is undistinguishable (except in being rather more prickly) from some specimens of *R. pampinosus* (Lees), which I place confidently under *R. villicaulis*.

β *derasus;* caule patenti-piloso setoso aculeis tenuibus patentibus vel paululum declinatis e basi compressa, foliis ternatis vel quinatis, foliolis tenuibus subtus in venis tantum pilosis, foliolo terminali late cordato-obovato cuspidato, panicula setosa.

R. derasus Müll.! Mon. 166 (1859).
R. vulgaris Lindl.! Syn. ed. 1. 93 (not W. and N.).
R. adsitus Genev.! (sp.)

Our plant is exceedingly like the authentic specimens of *R. derasus* preserved in Mr Baker's Herbarium. It is also determined to be the *R. vulgaris* of Lindley's first edition by a specimen so named by him for Leighton. It has much more setæ than I have ever seen upon even the most abnormal forms of *R. villicaulis*, and may very probably be a distinct species connecting the *Villicaules* with the *Radulæ* or even the *Bellardiani*. The tips of the long aciculiform setæ are deciduous, and then they may be easily confounded with the aciculiform prickles; but the proper setæ are very short, as are also the hairs. There are very few setæ (on our plant, but not on that of Müller) and many aciculi on the flowering shoot. Leighton's specimen was gathered at Almond Park, Salop; my specimens were found near Capel Curig in N. Wales, and Douglas, Isle of Man. Prof. Oliver found what is probably the same plant between Bonar Bridge and Lairg, Sutherlandshire, and Mr Bloxam sends it under the provisional name of *R. Bakeri* from Twycross, Leicestershire. If not a distinct species, and then it must

13

bear Müller's name *R. derasus*, it is probably a form of *R. villicaulis* when growing in damp situations. It also appears to be the *R. adsitus* (Genev.!) found by Mr Baker between Eastgate and Westgate in Weardale. The *R. vulgaris* of Lindley's ed. 2. is *R. Balfourianus.*

Habitat.—Woods and hedges. July, August.

Area.—1 2 3 . 5 . 7 8 . 10 11 12 . 14 15 . 17 . 19 20 . . 23 . . 26 . . . 30.

Localities.—i. Plympton St Mary, *S. Dev.* (Briggs!); Dunster, *W. Som.* (T. B. Flower); Leigh wood near Bristol, *N. Som.;* Tresco, Scilly, *W. Corn.* (F. Townsend!).—ii. Between Ryde and Quarr, *Isle of Wight;* Poole, *Dors.* (Bell Salter!).—iii. Mardley Heath, *Herts;* Pinner wood, *Middl.* (Hind!); Chertsey, *Surr.*—v. Wych and Forest of Dean, *W. Glouc.;* Redwood near Cheltenham, *E. Glouc.;* Newland, *Monm.;* Cowleigh Park, and Wire Forest, *Worc.;* Atherstone and Hartshill, *Warw.;* Seckley wood, Shawbury Heath and Almond Park, *Salop.*—vii. Cwm Bychan, *Merion.;* Llanberis, *Caern.;* Glan Hafren, *Montgom.*—viii. Bardon Hill and Packington, *Leic.*—x. Thirsk, *N. E. York.*—xi. Stannington and Barden Mills, *Northumb.*—xii. Douglas, *Isle of Man.*

xiv. *Linlithgow.*—xv. Campsie, *Stirl.* (G. E. Hunt).

xix. Killarney, *S. Kerry.*—xx. *Waterford.*—xxiii. *Meath* (D. Moore!).—xxvi. Maam, *W. Galw.*—xxx. Bushmills, *Antr.*

β—v. Almond Park, *Salop.*—vii. Capel Curig, *Caern.*— xvii. Between Bonar Bridge and Lairg, *E. Suth.* (D. Oliver!)

20. R. macrophyllus Weihe.

R. *caule arcuato-prostrato* angulato patenti-piloso, *aculeis* e basi magna dilatata compressa declinatis *brevibus tenuibus* conico-compressis, foliis quinatis, *foliolis duplicato-patenti-dentatis* vel irregulariter dentato-serratis supra pilosis subtus pallide viridibus tomentosis hirto-velutinis vel in venis tantum pilosis, foliolo terminali elliptico rotundo-obovato vel obovato cuspidato vel acuminato basi plus minusve cordato, paniculæ hirtæ tomentosæ setosæ ramis paucifloris corymbosis brevibus inferioribus axillaribus subracemosis ascendentibus, aculeis declinatis, sepalis hirtis tomentosis setosis ovato-attenuatis a fructu laxe reflexis.

R. macrophyllus Bab.! Man. ed. 5. 102 (1862); ed. 6. 111; Syme's Eng. Bot. iii. 177.

a *umbrosus* (Arrh.); aculeis e basi magna tenuibus, foliis quinatis, *foliolis duplicato-patenti-dentatis subtus hirto-velutinis plus minusvet omentosis*, foliolo terminali late obovato cuspidato, aculeis paniculæ tenuibus, *sepalorum apice lineari*, corolla rosea.

R. umbrosus Arrh. 31 (1840). Fries! Summa, 166; Herb. Norm. xiii. 60 (sp.). Bor. Fl. Centre, 200.
R. carpinifolius Blox.! in Kirby, 44. Lees! in Steele, 58.
R. carpinifolius β *umbrosus* Bab.! Man. ed. 6. 111.
R. Sprengelii Arrh.! in Fries, Herb. Norm. x. 53 (sp.).
R. atrocaulis Müll.! Mon. 90?

Stem forming a rather large arch at the base but with a long prostrate shoot beyond, nearly round and with slender scattered prickles at the base, hairy, having many subsessile glands, rather angular towards the top (even becoming slightly furrowed when dry). *Prickles* on the angles of the stem, slender, declining, from a large low oblong base. *Leaves* quinate. *Leaflets* convex, doubly dentate-serrate in a rather irregular manner, often with ascending serrate lobes in the upper half (traces of which may be seen in most cases, even when the serratures are nearly regular), dull green and pilose above, pale green, hairy on the veins and finely felted beneath or whiter with fewer hairs and more felt ; lower slightly stalked, obovate-lanceolate ; intermediate obovate, cuspidate, shortly stalked; terminal long-stalked, broadly obovate, cuspidate, often slightly cordate at the base ; petioles not furrowed, and as well as the midribs having strong hooked prickles beneath ; stipules narrowly linear-lanceolate.

Flowering shoot hairy, felted, glandular, from ashy scales. *Prickles* slender, long, declining, from a long compressed base. *Leaves* quinate or ternate. *Leaflets* very hairy on both sides, rather paler beneath, obovate, shortly cuspidate, toothed as on the stem ; basal leaflets of the quinate leaves subsessile, of the ternate slightly stalked unequal-based or lobate. *Panicle* long, narrow; branches short, few-flowered, corymbose, or 1-flowered, few lowest axillary, ascending, racemose ; rachis and peduncles hairy, felted, with many sunken setæ. *Sepals* lanceolate with a long flattened almost leaflike point, reflexed from the fruit, hairy, felted, setose, slightly aciculate. *Petals* contiguous, nearly white, clawed, roundish, toothed. *Filaments* pinkish. *Anthers* yellow. *Styles* cream-coloured, faintly pink at the base. *Primordial fruit-stalk* equalling the calyx.

This is certainly not the *R. carpinifolius* of many foreign botanists. The peculiar dentition of the leaves, which are usually hairy or even a little felted beneath, distinguishes it from the typical *R. macrophyllus;* and the slender prickles on the panicle, slightly felted leaves, and the shape of the terminal leaflet, usually separate it from *R. Schlechtendalii.*

Two specimens in my herbarium seem to belong to this variety of *R. macrophyllus,* but do not accord well with the characters given above. Both have a very much less angular and more hairy stem, bearing much more slender prickles. One is from Essendon, Herts, and was named *R. carpinifolius* by Bloxam (as I was informed by Coleman); and the other is from Gamlingay, Cambridgeshire, and in addition to those peculiarities has no felt and not much hair upon its leaves.

M. Genevier says that this is not *R. macrophyllus* (W. and N.), and in that opinion I quite agree with him, for it certainly is not the segregate species so named; nevertheless I still believe that it is properly combined with that plant and the others which I have grouped under my aggregate species thus denominated. He points out that a plant found at Marsden, Durham, by Mr Baker is the *R. atrocaulis* (Müll.), and refers it to *R. umbrosus.* It has the very small dentition of the doubtful plant found at Gormire near Thirsk, but is probably correctly placed here. I also place here the *R. flexicaulis* (Genev.!) which grows by Loch Awe (Hailstone!), in Birchin Grove near Worcester (Lees!), at Gamlingay in Cambridgeshire, and at Lyston in Herefordshire.

β *macrophyllus* (W. & N.); aculeis e basi maxima parvis brevibus, foliis quinatis vel ternatis, foliolis irregulariter dentato-serratis subtus infrequens tomentosis in venis pilosis, foliolo terminali elliptico vel late obo-

13—3

vato, aculeis paniculæ tenuibus, *sepalorum apice sæpe* foliaceo-*dilatato,* corolla alba.

R. macrophyllus Rubi Germ. 35. t. 12 (1825?). Borr.! in Eng. Bot. Suppl. t. 2625 ; in Hook. Brit. Fl. ed. 2. 246; ed. 3. 250. Leight.! Fasc. (sp.). Johnst.! E. Bord. 63. Lees, Malv. 56. Bor. Fl. Centr. 201. Wirtg.! Rub. Rhenan. Nos. 11, 79, 80 (sp.). Billot! Fl. Gall. et Germ. exsic. No. 1660 (sp.). Syme's Eng. Bot. iii. 177. t. 450.

R. vulgaris δ *macrophyllus* Sond. Hamb. 276.

R. vulgaris Leight.! Shrop. 231 (in part).

R. velutinus Weihe! in Reichenb. Fl. exsic. No. 785 (sp.).

R. Schlechtendalii Billot! Fl. Gall. et Germ. exsic. No. 1469 (sp.).

R. hispidus Merc. Cat. de Genève, teste Genevier!

Stem at first nearly upright, then curving down to the ground and extending itself close to the surface, angular, furrowed towards the end, having a variable quantity of short mostly patent deciduous hairs, and sometimes a few short setæ and aciculi, also rarely a little felt. (An authentic specimen of the plant figured in *English Botany* has a considerable quantity of felt on its stem.) *Prickles* usually few, declining, short, conical, but a little compressed, rather slender, often shorter than and rarely longer than the greater diameter of their very long compressed low bases. *Leaves* quinate or ternate, subpedate. *Leaflets* rather thin, green on both sides, with scattered hairs above, paler and hairy on the veins beneath (the spaces between the veins being either quite naked and rough, or more or less densely felted), rather irregularly dentate or doubly patently dentate; basal shortly-stalked, oblong, acute ; intermediate obovate, subcuspidate or subacuminate, often subcordate at the base ; terminal long-stalked, very variable in shape from roundly obovate to very long cuneate-obovate, acuminate or subcuspidate, usually more or less cordate at the base ; midribs and petioles with slender hooked prickles beneath.

Flowering shoot from brown silky scales, angular, hairy, nearly without setæ or aciculi. *Prickles* from a long compressed base, slender, declining, or rather strong. *Leaves* mostly ternate. *Leaflets* varying like those of the stem, pilose above, paler and hairy on the veins beneath, dentate-serrate or sublobate-dentate towards their tip ; basal usually unequal-sided ; uppermost floral leaves sometimes simple, three-lobed, or ternate, with the terminal leaflet wedge-shaped at the base and very shortly stalked. *Panicle* short, with two or three axillary subracemose ascending branches which fall short of the leaves ; rachis and peduncles hairy, felted, with many yellow subsessile glands, and a few aciculi and short purple setæ. *Sepals* ovate-attenuate, with a slender leaflike or flat and linear point, hairy, felted, very slightly setose, loosely reflexed from the fruit. *Petals* oblong, white. *Stamens* white. *Styles* cream-coloured. *Nuts* ovate ; inner edge straight.

R. macrophyllus, even as restricted in the *Rubi Germanici*, is a very variable plant. The terminal leaflet is sometimes nearly circular, cuspidate, and scarcely at all cordate at the base ; but a series of plants may be found connecting that form of leaf with one which is cordate-acuminate or cordate-obovate-acuminate. In all of them the dentition is very nearly simple and regular. The stems of *R. macrophyllus* are often furnished with a few aciculi and setæ, which are usually short and have thick bases ; but very rarely a plant clearly belonging to *R. macrophyllus* is found to have almost as great an abundance of those minute arms as the species of the Section *Radulæ*. They prove to us that our sections are not so clearly defined in nature as in our arrangements.

γ *Schlechtendalii* (W. & N.); aculeis e basi maxima parvis brevibus, foliis sæpissime quinatis, *foliolis dupli-cato-patenti-dentatis* subtus sæpissime in venis tantum pilosis nec tomentosis, *foliolo terminali longe obovato* acuminato *basi cuneato* vel subcordato, aculeis paniculæ validis, sepalorum apice lineari, corolla alba.

R. macrophyllus β *Schlechtendalii* Bab.! Syn. 20. Leight.! Fasc. (sp.).

R. Schlechtendalii Rubi Germ. 34. t. 11 (1825?). Blox. in Kirby 45.

R. vulgaris γ *Schlechtendalii* Sond. Hamb. 276.

R. piletostachys Godr.! in Gren. et Godr. Fl. de Fr. i. 548; Fl. Lorr. ed. 2. i. 242. Billot! Fl. Gall. et Germ. exsic. No. 2667 (sp.). Wirtg.! Herb. Rub. ed. 1. No. 131 (sp.).

R. mentitus Müll.! in Billot Annot. 293.

R. macrophyllus var. Wirtg.! Herb. Rub. 79 and 11 b. (sp.).

The difference between this plant and the *R. macro-phyllus* is so slight that a detailed description is unnecessary; indeed the characters supposed to separate them are incon-stant. It sometimes happens, as is remarked in the *Rubi Germanici*, that the leaves on the flowering shoot are felted on the underside, whilst those on the stem are nearly or quite devoid of felt.

The *R. piletostachys* (Godr.) has a broader terminal leaflet than is usual on this plant; and it is decidedly cordate at the base. I should describe it as roundly-quadrangular-obo-vate acuminate-cuspidate, coarsely and doubly and patently dentate. *R. piletostachys* seems to connect this plant with ε *glabratus;* but perhaps it may be really distinct; for Godron says that the stem is erect-arcuate and the petioles flat above.

A specimen from Bloxam (unnamed), from Hartshill wood, has a very long lax panicle and remarkably slender but exceedingly long-based prickles on both shoots. It probably is a state of this variety. It is in the Herb. Borrer.

δ *amplificatus* (Lees); aculeis e basi maxima brevibus, foliis sæpissime quinatis, *foliolis* subduplicato-patenti-dentatis *subtus in venis tantum pilosis* nec tomentosis, foliolo terminali late obovato acuminato, paniculæ maximæ aculeis e basi maxima compressa validis, sepalorum apice sæpe foliaceo-dilatato, corolla alba vel subrosea.

R. amplificatus Lees! in Steele 58 (1847); Malv. 56. Blox.! in Kirby 45.

R. macrophyllus γ *amplificatus* Bab.! Syn. 20; Man. ed. 2. 101; ed. 6. 111.

R. umbraticus P. J. Müll.! in Wirtg. Herb. Rub. No. 82 (sp.).

R. stereacanthus P. J. Müll. (teste Genevier).

Very nearly allied to *R. Schlechtendalii*. *Leaflets* hairy only on the veins beneath, rather irregularly dentate with the larger teeth somewhat reflexed; terminal roundly obovate, acuminate, sometimes subcordate at the base; petioles and midribs with strong but slender hooked prickles beneath.

Flowering shoot with very strong (especially in the panicle) deflexed or declining prickles springing from exceedingly long compressed bases. *Panicle* leafy, long; branches mostly axillary, lower forming secondary panicles, upper short racemose-corymbose exceeded by the leaves. *Sepals* with a slender leaflike point. *Petals* rather distant, obovate, toothed, mostly white but sometimes tinged with purple. *Filaments* white. *Anthers* and *styles* greenish.

ε *glabratus;* aculeis e basi maxima brevibus, foliis quinatis, *foliolis* irregulariter vel subduplicato-dentatis *subtus in venis tantum sparsim pilosis, foliolo terminali cordato-subrotundo* vel late obovato basi subcordato, aculeis paniculæ tenuibus, sepalorum apice sæpe foli-aceo-dilatato (?).

R. *vulgaris* γ *glabratus* Rubi Germ. 38. t. 14. δ? Bab. Man. ed. 6. 111.

This variety differs chiefly from *var.* β and γ by the nearly glabrous underside of its leaves and the remarkably round form of the terminal leaflet, which is usually, but not always, cuspidate. I have very little acquaintance with it, and derive almost all my knowledge from a series of specimens kindly sent to me by Mr H. C. Watson, who gathered them near Long Ditton, Surrey. Two of these specimens were named R. *cordifolius* by Mr Bloxam (in 1853), but they do not agree with authentic specimens of that plant.

Some doubt attends the identification of our plant with that of Germany, for the leaves of the latter are said to be soft beneath.

Careful consideration and the examination of many specimens has led me to the conclusion that all these plants are probably forms of one variable species, notwithstanding the rather considerable differences which exist between well developed states of them. I am pleased to find myself con-firmed in this opinion by a botanist of such eminence as Mr Sonder (*Fl. Hamb.* 275); although he adds to the group, as I think erroneously, the R. *carpinifolius* of the *Rubi Ger-manici.* It is nevertheless quite possible that the error may rest with us, and that the British R. *carpinifolius* is different from that of Germany. The habit of our plant seems to keep it quite separate from any form of R. *macrophyllus.* R. *umbrosus* and R. *Schlechtendalii* are usually well marked by the peculiar dentition of their leaves, but traces of a similar structure may occasionally be seen in plants of *R. macrophyllus* proper.

The leaflets with a felted underside and also hairs upon the veins of *R. umbrosus* and *R. macrophyllus* proper, generally seem very different from those of *R. Schlechtendalii* (which are usually totally devoid of felt and only bear a quantity of long hairs on the veins); but it sometimes happens that a very thin coat of felt may be seen even upon the latter by using a glass of strong magnifying power. The shape of the terminal leaflet is inconstant: that of *R. Schlechtendalii* is usually very long and wedgeshaped even to the extent of the lower half or two thirds of its length: in *R. macrophyllus* it is generally much shorter in proportion but always apparently rather broader above than below its middle. The leaves of the latter are frequently not more than ternate, either simply or with lobed lower leaflets. The panicle of *R. umbrosus* will usually distinguish it from the other forms. It is pretty constantly narrow, long, its upper part leafless through some extent with patent short branches: even the lower branches also are sometimes patent. The other plants have, normally, short subcorymbose panicles of which all the branches ascend. The presence of intermediate states of panicle, and different combinations between them and the form of the leaves, shows that they do not characterise species in the present case. Metsch combines *R. umbrosus* with *R. carpinifolius* and *R. villicaulis* under the name of *R. vulgaris* (Weihe). Certainly their panicles are often very much alike; but that, I think, is their chief point of resemblance.

The *R. amplificatus* is well marked by the very strong prickles upon its panicle, which is long with many distant mostly axillary branches, and of these the lower are often very long. The plant from Great Cowleigh Park, called

R. Babingtonii in my *Synopsis*, is *R. amplificatus*. Mr Lees considers (1864) *R. amplificatus* to be quite distinct from *R. macrophyllus*, and to be known by "its almost smooth stem, leaves green on both sides and generally short corymbs."

The *R. cordifolius* of Johnston's *East. Borders* is apparently a form of *R. macrophyllus*, but does not accord well with either of the described varieties. Its leaflets are not at all felted beneath, the terminal leaflet is nearly round and cuspidate; the panicle has long axillary lower branches. The prickles of both shoots and of the panicle are slender, but spring from long compressed bases. An aciculus or seta may be found occasionally on the stem. Setæ are more abundant amongst the felt and hairs of the panicle, and are usually long and prominent.

I have received several specimens of this species from M. Questier. Those named *R. macrophyllus* and *R. Schlechtendalii* belong to my var. *β*. His *R. vulgaris* var. *umbrosus* is the same as my var. *a*, but has much less long hair on the underside of the leaves, but there is a coat of very short and thin felt. ·

Habitat.—Woods and thickets. July, August.

Area.—1 2 3 4 5 6 7 8 9 10 11 12 13 14 15 16 . . 19 . . . 23 . . . . . . 30.

Localities of var. *a umbrosus.*—i. Liskeard, *E. Corn.;* Heale, *N. Devon.*—ii. St Leonard's Forest, *W. Suss.* (Borr.!). —iii. Essendon, *Herts.*—iv. Sandy, *Beds.;* Dunwich, *E. Suff.*—v. Llanrumney, *Monm.;* Forest of Dean, *W. Glouc.* (Hort.!); Malvern Hills, *Heref.* and *Worc.;* Haughmond Hill, *Salop.*—vi. *Cardigan.*—vii. Llanberis, *Caern.;* Hen blas Cromlech, *Angl.;* Capel Garmon, *Denb.;* Cwm Bychan, *Merion.* (Borr.!)—viii. Twycross, *Leic.;* Chalk Abbey, *Derby* (Bloxam).—ix. Manchester (Sidebottom!), *S. Lanc.*—x. Bell Hag near Sheffield, *S. W. York.;* Symmingthwaite, *W. York.*

(Baker!).—xi. Alnwick, *Northumb.* (Baker).—xii. Skelwith, *Westm.*; Derwentwater, *Cumb.*

xiii. Jardine Hall, *Dumf.*; Gourock, *Renf.*—xvi. Lamlash in Arran, and Bute (Balfour!), *Clyde Isles;* Islay, *S. Ebudes.*

Of β *macrophyllus.*—i. Near Plymouth, *S. Dev.* (Briggs); Warmscombe, *N. Dev.*; Culbone, *S. Som.*—ii. Henfield, *W. Suss.*—v. Rogerstones, *Monm.*; Almond Park, *Salop.*—vi. Aberystwith, *Card.*; Stackpole, *Pemb.*—vii. North of Dolgelly, *Merion.*; Llanberis, *Caern.*—viii. Higham, *Leic.* (Bloxam!); Matlock, *Derby* (Backhouse).—xi. Knutsford, *Chesh.*—xi. Twizel Dean, Ancroft, and Haggerstone, *Chev.* (Johnston!).—xii. Bowness and Ambleside, *Westm.*

xiv. Hirsel Law, *Berw.* (Johnston!).—xv. By the river Don at Aberdeen, *S. Aberd.*; Campsie, *Stirl.* (G. E. Hunt!).

xix. Killarney, *S. Kerry.*—xxx. Londonderry, *Derry* (D. Moore!); Black mount (Hind!), Laganside, *Antrim* (Tate!).

Of γ *Schlechtendalii.*—i. Calstock, *E. Corn.*; Torquay, and Elburton, *S. Dev.*; Chambercombe, *N. Dev.*—ii. Henfield, *W. Suss.* (Borr.!).—iii. Claygate, *Surr.*; Welwyn, *Herts.*—v. Near Chepstow, *Monm.*; Malvern, *Worc.*; Shrewsbury, *Salop.*—vi. Aberystwith, *Card.*—vii. Llanberis, *Caern.*—viii. Twycross, *Leic.* (Blox.!).—x. Thirsk, *N. E. York.*

xiv. Winchburgh, *Linlith.*—xv. Inverness, *Eastern.*

xix. Killarney, *S. Kerry.*—xxiii. *Wicklow.*

Of δ *amplificatus.*—v. Great Cowleigh near Malvern, *Worc.*—viii. Twycross, *Leic.*; Chalk Abbey, *Derby* (Bloxam).—x. Hooton Cliff, *S. W. York.* (Bloxam!); Boltby, *N. E. York.*

xvi. Lamlash in Arran, *Clyde Isles.*

xxx. Carumoney, *Antrim* (Tate!).

Mr Lees considers this as the most common of the plants I include under *R. macrophyllus;* so apparently does Mr Syme. I record only those localities concerning which I have certain information.

Of ε *glabratus.*—iii. Long Ditton, *Surr.*—v. Almond Park, *Salop.*—vii. Llanberis, *Caern.*—ix. Knutsford, *Chesh.* xix. Killarney, *S. Kerry.*

c. Spectabiles. Aculei caulis plus minusve inæquales; setæ et aciculi breves perpauci, pili sæpe densissimi.

It is an exception to the rule for brambles belonging to the *Sylvatici* to have setæ and aciculi: but the rule is for the *Spectabiles* to possess them.   M. Müller, who formed these groups, has failed in pointing out any good and constant distinctions between them, nor have I been more successful. Nevertheless the groups seem to be natural.

21. R. mucronulatus Bor.

R. caule arcuato *subtereti* patenti-piloso, *aculeis* paucis e basi oblonga dilatata conicis *tenuibus* declinatis, foliis quinatis, *foliolis crassis* argute *dentato-serratis* utrinque viridibus rugosis supra pilosis subtus in venis (sæpissime rufescentibus) tantum hirtis, *foliolo terminali* late obovato-*cuspidato* basi cordato, paniculæ angustæ foliosæ laxæ pilosæ tomentosæ setosæ ramis longis 1-3-floris aculeis paucis tenuibus declinatis, sepalis hirtis tomentosis setosis laxe reflexis ovate-attenuatis apice lineari.

R. mucronulatus Bor. Fl. Centre, ed. 3. 196 (1857). Bab.! Man. ed. 5. 103; ed. 6. 112. Syme's Eng. Bot. iii. 178. t. 451.

R. mucronatus Blox.! in Kirby, 43 (1850). Bab.! in A. N. H. ser. 2. ix. 126; Man. ed. 3. 97; ed. 4. 100. Johnst.! East. Bor. 66 (not of Ser. in DC. Prod. ii. 565).

R. sylvaticus Bab.! Syn. 16 (in part). Blox.! Fasc. (sp.). Leight.! Shrop. Rubi (sp.).

R. vulgaris Lindl.! Syn. ed. 2. 93 (in part). Leight.! Fl. Shrop. 231 (in part).

R. leucanthemus Müll. Mon. 49 (1859) (teste Genevier!).

R. amplichloros Müll. in Boulay, Ronces des Vosges, 10 (1859) (teste Genevier).

R. Lingua Lees in Steele, 57.

Stem arched, nearly round, slightly angular with flat sides or slightly furrowed towards the end, densely hairy near the base but less so towards the end, often becoming

nearly naked; hairs patent, not clustered; aciculi and setæ
few or none; subsessile glands few. *Prickles* chiefly on the
angles of the stem, few, usually small, slender, conical from
an enlarged base, patent or very slightly declining. *Leaves*
quinate. *Leaflets* rather thick, dark green, rough and pilose
on both sides, pale with more numerous hairs on the veins
beneath, finely dentate-serrate, nearly flat with the edge
slightly turned upwards; lower shortly stalked, obovate-
oblong, cuspidate; intermediate larger, stalked, obovate,
abruptly cuspidate; terminal with a rather long stalk,
broadly obovate with a cordate base, abruptly cuspidate;
petioles and midribs with few small deflexed prickles beneath; stipules linear-lanceolate.

Flowering shoot long, from fuscous scales, slightly angular,
green but tinged with purple, hairy. *Prickles* few, generally
very small and short, yellow, sometimes long, straight and
declining, slender, from an enlarged and compressed base.
Leaves ternate or quinate. *Leaflets* nearly equally hairy on
both sides, rather paler beneath; of the ternate leaves nearly
equal, oblong or obovate, finely serrate, lower often lobed
externally; of the quinate leaves the lower leaflets are small
and oblong, intermediate and terminal broadly obovate and
cuspidate; petioles and midribs with few slender declining
prickles beneath; stipules linear-lanceolate. *Panicle* narrow,
very lax, leafy except at the top, hairy and felted, often with
many unequal red setæ and aciculi; branches mostly axillary,
ascending, falling short of the leaves, bearing a corymb of
1-3 long-stalked flowers; summit corymbose. *Sepals* lanceolate, acuminate, with a long linear point, hairy, felted,
setose, and greenish, with a narrow margin of white felt
externally, whitely felted but purple at the base within,
loosely reflexed from the fruit. *Petals* distant, obovate,
clawed, pale pink, entire. *Filaments* pink at the base.
Anthers greenish. *Styles* cream-coloured, pink on the young

fruit. *Primordial fruit* small, hemispherical; its stalk rather long.

There is a specimen of what appears to be a very much developed form of this species in the *Herb. Borr.*, from "Bridge of Ogwan," Caernarvonshire. The stem and leaves are badly represented, but the panicle is magnificent. It is only part of the flowering shoot, but is 16 inches long and all panicle. The lower axillary branches fall short of the leaves, and bear, upon a long simple base, many-flowered corymbose cymes: the ultra-axillary part is broad, convex and densely-flowered. The upper floral leaves are almost exactly cordate. The panicle is very much more thickly covered with patent hairs than is usual; so thickly as almost completely to hide the few setæ that are amongst them. But notwithstanding the great difference that seems to exist between this fine panicle and the very nearly simple raceme of one of Bloxam's specimens from Hartshill, and of my own from Islay, it is clear that their real structure is identical. In the ordinary state there are simple peduncles or corymbose cymes of very few long-stalked flowers; but when something has caused the cymose structure to be more fully developed we have the many-flowered cymes of the plant from Ogwan.

Mr Lees informs us (*Bot. of Worc.* 42) that this was his former *R. Lingua.*

M. Genevier has combined several plants distinguished by P. J. Müller with this species. They seem to differ from it at the first view, but are probably not deserving of separation from *R. mucronulatus.* They are the *R. leucanthemus* of Müller's *Monograph*, and the *R. amphichloros* described by him in M. Boulay's *Ronces des Vosges*, a book which I have not seen.

As this is a well-defined species no further remark is requisite. I have received it from M. Questier as a possible form of *R. Babingtonii.* Unfortunately there is a bramble from Newfoundland to which Seringe gave the name of *R. mucronatus* in 1826 in De Candolle's *Prodromus* (ii. 565). Boreau has therefore slightly altered the name of our plant.

Habitat.—Hedges and banks. July, August.

Area.—1 . . . 5 . 7 8 . 10 . . . 14 15 16.

Localities.—i. Near Plymouth, *S. Dev.* (Briggs!).—v. Hartshill wood, *Warw.;* Shawbury Heath, *Salop.*—vii. Ogwan Bridge, *Caern.* (Borr.!).—viii. Seale wood near Twycross, *Leic.* (Blox.).—x. Thirsk and Laskill, *N. E. York.*—xi. Hartley, *Northumb.*

xiv. Rare in *Berwickshire* (Johnston!); Winchburgh, *Linlithg.*—xv. Rubieslaw, *S. Aberd.;* Campsie, *Stirl.* (G. E. Hunt!).—xvi. Islay, *S. Ebudes;* Killmalie by Loch Eil, *Western.;* Lamlash, *Arran* (Balfour!).

22. R. Sprengelii Weihe.

R. caule prostrato piloso, *aculeis inæqualibus* e basi magna compressa *deflexis*, foliis 3-5-natis pedatis, foliolis tenuibus utrinque viridibus subtus in venis sparsim pilosis, foliolo terminali elliptico-acuminato, paniculæ laxæ hirtæ tomentosæ setosæ ramis axillaribus patentibus paucifloris summis extra-axillaribus divaricatis, aculeis paucis tenuibus deflexis, sepalis ovatis acuminatis erecto-patentibus apice sæpe foliaceo-dilatato.

R. Sprengelii Sond.! Hamb. 275. Bab.! Man. ed. 3. 98; ed. 6. 112. Blox.! in Kirby, 44. Lees in Steele, 58. Godr. in Fl. de Fr. i. 542. Bor. Fl. Centre, 201. Metsch in Linnæa, xxviii. 156. Syme's Eng. Bot. iii. 179.

a Borreri; caule arcuato-procumbente crasso sparsim aciculato et setoso, aculeis inæqualibus, foliis sæpe quinatis, panicula subthyrsoidea vel ad apicem subcorymbosa.

R. Sprengelii a Borreri Bab.! Man. ed. 3. 98; ed. 6. 112.
R. Borreri Bell Salt.! in A. N. H. xv. 306 (1845). Bab.! Syn. 17; Man. ed. 2. 100.
R. Sprengelii Lange! Danske Fl. ed. 2. 347. Reichenb.! Fl. exsic. 784 (sp.). Wirtg.! Herb. Rub. No. 51 (sp.). Billot! Fl. Gall. et Germ. exsic. No. 971 (sp.).

Stem usually lying close to the ground or very slightly arching, thick, often terete, with a few setæ and hairs and a very few short large-based aciculi. *Prickles* many, not confined to the angles, very unequal, often much deflexed, conical, from a long compressed base. *Leaves* quinate-pedate or ternate. *Leaflets* all stalked, rather thin, similarly green on both sides, distantly pilose on the veins

beneath, rather irregularly serrate; basal and intermediate
lanceolate; terminal shortly and broadly obovate, acuminate;
or the ternate leaves with elliptic-acuminate rather unequal-
sided basal leaflets; midrib and petiole, which is not furrowed above, with few strongly hooked prickles beneath;
stipules linear-lanceolate.

Flowering shoot from fuscous scales, very hairy. *Prickles*
few, small, strong, from a long base. *Leaves* ternate.
Leaflets elliptical, acute at both ends. *Panicle* corymbose
or thyrsoid, slightly leafy below; branches divaricate,
corymbose, few (2-4) flowered, axillary ones patent and
falling short of the leaves, or rather pyramidal with a subcorymbose top; lower branches rather distant, axillary,
longer than the others but not equalling the leaves; rachis
and peduncles with few slender slightly declining prickles,
few aciculi, more numerous sunken setæ. much hair and felt.
Sepals ovate-acuminate, leaf-pointed, hairy, felted, with many
short sunken setæ, erect-patent and slightly clasping the
fruit. *Petals* narrow, obovate, entire, clawed, pink. *Filaments* pink. *Anthers* greenish. *Styles* pale green. *Primordial
fruit* hemispherical; stalk shorter than the sepals. *Nut*
½-ovate; inner edge nearly straight.

β *Sprengelii;* caule sæpissime prostrato tenui, aciculis et setis subnullis, aculeis parvis, foliis sæpe ternatis, foliolis flexibilibus, panicula laxa pauciflora sub-
corymboso-pyramidata.

R. *Sprengelii* β *Sprengelii* Bab.! Man. ed. 3. 98; ed.
6. 112.

R. *Sprengelii* Weihe in Bot. Z. (Flora) ann. 2. ii. 18
(1819); in Rubi Germ. 32. t. 10. Tratten. Ros. iii. 39.
Bab.! Man. ed. 2. 100; Syn. 17. Wirtg.! Rub. rhenan. 51
(sp.). Fr.! Herb. Norm. xv. 49. (sp.). Genevier (sp.!).

R. Arrhenii Lange! Danske Fl. ed. 2. 347 (1859).

R. saxatilis Reichenb.! Fl. Germ. exsic. No. 787 (sp.).

Stem prostrate, round, slender, angular towards the end, slightly hairy. *Prickles* many, strongly deflexed, from a large base. *Leaves* ternate or quinate-pedate. *Leaflets* irregularly serrate, flexible, bright shining green with a few hairs above, green and pilose on the veins beneath; all lanceolate or slightly obovate acuminate; basal of the ternate leaves usually strongly lobed below; intermediate of quinate leaves unequal-based; petioles which are furrowed above and midribs with very few small prickles beneath; stipules linear-lanceolate.

Flowering shoot from brown rather silky scales, hairy, with usually few deflexed prickles, all seeming to be radical. *Leaves* ternate. *Leaflets* often rather strongly serrate, more hairy on both sides than those of the stem; basal broadly lanceolate; terminal broadly obovate, acuminate. *Panicle* broad, short, hairy, setose; lower branches axillary, erect-patent, short, few-flowered; upper divaricate, often 1-flowered. *Sepals* ovate, acuminate, leaf-pointed, hairy, felted, setose, clasping the fruit. *Petals* distant, obovate, acute, entire, pink. *Filaments* pinkish. *Anthers* and *styles* greenish. *Primordial fruit-stalk* as long as the sepals.

There cannot be any doubt that the *R. Sprengelii* (Weihe) is a small form of the species of which *R. Borreri* is the type. The species does not present much difficulty to the student when thus considered. It is unfortunate that the law of priority obliges us to adopt the name given to a mere form as that of the species. It is justly remarked by Sonder that the former *R. Sprengelii* of Arrhenius! and Fries! is quite different from that of Weihe. It seems to be *R. macrophyllus a umbrosus*.

I do not possess a specimen of the *R. rubricolor* of

Bloxam which is described by Mr Syme (*Eng. Bot.* l. c. 180), but judging from the description given by him I suppose it to be an extreme state of *R. Sprengelii* or rather of *R. Borreri.* Mr Syme's words are "Barren stem arching, very stout, prickles numerous, nearly destitute of gland-tipped setæ and aciculi. Leaves quinate ; terminal leaflet oblong obovate-cuspidate. Panicle lax, many-flowered ; rachis more densely setose, with numerous strong prickles and a few aciculi and gland-tipped setæ." Mr Bloxam finds it near Mansetter, Warwickshire, and considers it as a distinct species, and the same as *R. erubescens* (Wirtg.).

The *R. Borreri* of Billot's *Fl. Gall. et German. exsiccata* (No. 1867) is not exactly either of our forms of *R. Sprengelii.* It seems to me that it is nearer to the original *R. Sprengelii* than to our *R. Borreri.* No. 971 of the same valuable collection is named *R. Sprengelii,* and is exactly the *R. Borreri* of Bell Salter. It is manifest from this that at least some Continental botanists regard the species precisely as I do.

Habitat.—Woods and heaths. June, July.

Area.—. 2 3 . 5 . 7 8 9 10 . 12.

Localities of *a.*—ii. Niton, *I. of W.*—v. Newchurch, *Monm.;* Coleford, *W. Glouc.;* Rugby and Atherstone, *Warw.* —vii. Pass of Llanberis, *Caern.*—viii. Southwood and Chalk Abbey, *Derby;* Coleorton, *Leic.* (Blox).—x. Sheffield and Hebden Bridge, *S. W. York.;* Harrogate, *Mid. W. York.*— ix. Accrington, *S. Lanc.*

Of *β.*—iii. Hatfield, *Herts.*—v. Bromsgrove Lickey, *Worc.;* Newchurch, *Monm.*—vii. Rhaiader Mawddach, *Merion.*— viii. Bardon Hill, *Leic.*—ix. Bowdon (G. E. Hunt!), and Congleton, *Ches.;* Bredbury Wood near Manchester (Bloxam), Mere Clough near Prestwick (G. E. Hunt!), and Accrington, *S. Lanc.*—xii. Ambleside, *Westm.*

d. Radulæ. Caules punctis elevatis rigidis, ubi setæ aciculique breves subæquales sederunt, asperi efficiuntur; aculei subæquales.

The plants contained in this group have much in common. They may be known from the *Spectabiles* by having an abundance of short and equal aciculi and setæ on their barren stems. When such arms are found on *Spectabiles* they are inconspicuous, few, and scattered very irregularly; some internodes bearing a considerable number, but other parts of the stem totally wanting them. Here they are tolerably uniformly and universally distributed.

The *Glandulosi* differ from these plants by having very scattered prickles, which vary greatly in size, and decrease gradually and insensibly so as to become undistinguishable from aciculi, and the aciculi are similarly undistinguishable from the setæ and hairs. Their largest prickles are not confined to the angles of the stem. All these arms are persistent, and therefore the old stems of the *Glandulosi* are never rough in the same way as those of the *Radulæ*. Even in those cases where they make the nearest approach to the roughness of the *Radulæ*, a careful examination will show that the prominent points are not tubercles, but the somewhat cylindrical bases of broken and rather strong aciculi and setæ. The roughness of the stems of *Radulæ* arises from the permanent rather hemispherical bases of weak aciculi and setæ.

The markedly felted underside of the leaves separates
R. rudis and *R. Radula* from the other species included in
this group, except *R. scaber;* and from that plant their
highly arching stem distinguishes them.

R. Bloxamii is probably best known by the remarkably
large size and round form of its leaflets, even upon the
flowering shoot; by the panicle being leafy nearly to its top,
and its axillary branches being corymbose.

R. rosaceus shows an approach to the *Koehleriani* by
having a much less marked interval between its prickles,
aciculi, and setæ, than any of the other *Radulæ.*

The adpressed sepals, combined with a rather narrow
panicle and a very marked interval between the prickles and
other arms of the stem characterize *R. Hystrix.*

23. R. Bloxamii Lees.

R. caule arcuato-prostrato angulato subsulcato, *aculeis parvis* inæqualibus *subpatentibus,* aciculis setisque brevibus æque ac pilis multis, foliis 5-3-natis, *foliolis grosse* duplicato-*dentatis utrinque viridibus et pilosis, foliolo terminali rotundo-obovato* cuspidato basi subcordato, *paniculæ* longæ *usque ad apicem foliosæ* tomentosæ cum brevibus *ramis* et *apice corymbosis,* aculeis tenuibus declinatis, sepalis ovato-acuminatis a fructu laxe reflexis.

R. Bloxamii Lees! in Steele, 55 (1847). Bab.! Man. ed. 4. 101; ed. 6. 112.

R. Babingtonii β *Bloxamii* Bab.! in A. N. H. xvii. 244 (1846); Man. ed. 2. 102; ed. 3. 99. Blox.! in Kirby, 40.

R. Guntheri β *Bloxamii* Bell Salt. in Bot. Gaz. ii. 126 (1850); in Hook. and Arn. Br. Fl. ed. 7. 128.

R. rhenanus Müll. (teste Genev.).

Stem arcuate-prostrate, angular throughout, sometimes slightly furrowed. *Aciculi* and *setæ* many, equal, very short, deciduous, with the exception of their very thick bases. *Hairs* about equalling the setæ. *Prickles* many, unequal, small, conical, slightly declining from large compressed bases. *Leaves* quinate or ternate. *Leaflets* all stalked, dark green and distantly pilose above, paler green and hairy on the veins but not felted beneath, coarsely serrate near the base, doubly and almost lobate-serrate in the upper part; basal oblong, acute; intermediate broadly obovate-cuspidate, or obovate-lanceolate-acuminate; terminal roundly-obovate-cuspidate, or almost round with a cordate base (I have seen

15

one leaf where the terminal leaflet is lobed on one side of its
base and has thrown off a distinct oblong leaflet on the other
side); or the leaf becomes ternate by the combination of the
basal and intermediate leaflets; petioles, which are flat or
slightly furrowed above, and midribs with slender deflexed
prickles beneath; stipules very slender, linear-lanceolate.

Flowering shoot long, leafy, with many hairs setæ and
aciculi, and many slender declining prickles from large compressed bases.   *Leaves* usually ternate, like those of the stem
in all other respects, rarely subquinate; floral leaves very
large, uppermost simple, cordate and lobed or cordate-ovate.
Panicle long, usually leafy to the top, hairy, setose and
aciculate; branches many, distant, short, corymbose, few-
flowered, axillary, erect-patent; top corymbose; prickles
many, very unequal, slender, declining. *Sepals* shortly
ovate, abruptly acuminate, often with a long leaf-like point.
Petals rather distant, ovate, blunt, clawed, entire, white.
Filaments white.   *An'hers* and *styles* greenish.

Mr Bloxam long since stated it to be his opinion that
this plant did not associate well with *R. Babingtonii* (*R.
scaber*), and I have for some time fully agreed with him.   In
many respects it very nearly approaches the *R. thyrsiflorus*
of Weihe, the chief distinctions of which from *R. Bloxamii*
consist in its rounder stem, serrate rather than dentate
leaflets, and especially in the greater part of its panicle
being leafless, dense and cylindrical, and the branches (even
the two or three axillary ones) racemose.   These branches are
described as far nearer corymbose than they are represented
on the plate of the *Rubi Germanici*.   If the *R. thyrsiflorus* and .
R. Bloxamii vary in that respect, as seems probable from
this discrepancy in the *Rubi Germanici*, and from the fact
that of two specimens of the latter plant received from Mr
Leighton (who gathered them at Almond Park, near Shrewsbury) one has a more naked top to its panicle than is found

in the original *R. Bloxamii*, and the other has several race-mose branches; then the plants may ultimately prove to be identical. Unfortunately I have not seen authentic speci-mens of *R. thyrsiflorus*. The plant so-named by M. Questier approaches very nearly to *R. Bloxamii*, especially to one of the above-mentioned specimens from Mr Leighton.

A very beautiful plant, gathered at Kenilworth by Mr T. Kirk in 1854, closely resembles the figure of *R. thyrsi-florus* in the *Rubi Germanici* (tab. 34). Its panicle accords almost exactly with that plate, and must have been quite as large; but the sepals are rather loosely adpressed to the fruit, whilst those of *R. thyrsiflorus* are expressly stated to be reflexed from it. The leaves of the barren stem, as far as I know them, are very much smaller and more finely (although similarly) toothed. The stem has moderate-sized, compressed, declining, scattered prickles arising from very long bases, and an abundance of short rather unequal aciculi and setæ, most of them also springing from enormous bases. All my knowledge of this plant being derived from one specimen, it is out of my power to form a satisfactory opinion concerning it; but I am inclined to think that it is more nearly allied to *R. Bloxamii* than to any other bramble which is known to me. As I have not this specimen now before me I am unable to say how nearly it resembles the specimens shown to me by Mr Baker, gathered be-tween Eastgate and Westgate in Weardale, Durham. They are very near this species, even if they should not be joined with it. Genevier considers them as closely allied to *R. adscitus* (Genev.), but more prickly. It is probable therefore that Genevier's plant is very closely allied to *R. Bloxamii*, from which these specimens from Mr Baker seem chiefly to differ by their much fewer and more deciduous setæ, and much more naked panicle, of which not more than the few lowest branches are axillary. *R. adscitus* was described in

Genevier's *Essai* (Mém. Soc. Angers, viii.) in 1860; it was placed with *R. rosaceus* by Boreau, but is not the plant so named in the *Rubi Germanici*. It is not improbable that this plant and that of Mr Kirk may be the true *R. thyrsiflorus* (W. and N.), and may require to be separated from *R. Bloxamii*, and that *R. adscitus* will have to be combined with that species.

The *R. thyrsiflorus* of Boreau (*Fl. du centre de la France*, 195) seems to agree quite as well with our *R. Bloxamii* as does his *R. Bloxamii* described on the same page; if not indeed better.

Habitat.—Woods.   July and August.

Area.—1 ... 5 .. 8 .. 11 .... 30.

Localities. — i. Crabtree, *S. Dev.* (Briggs!). —v. Near Atherstone and by Hartshill wood, *Warw.;* Almond Park, *Salop.*—viii. Orton wood near Twycross, *Leic.* (Blox.).— xi. Weardale, *Durh. ?*

xxx. Black Mount near Belfast, *Antr.*

24. R. Hystrix Weihe.

R. caule arcuato-prostrato angulato subsulcato, aculeis e basi dilatato-compressa tenuibus declinatis aciculos pilosque paucos et etiam setas multas omnes inter se subæquales brevesque longe excedentibus, foliis quinato-pedatis, *foliolis grosse et subduplicato-patentidentatis utrinque viridibus et pilosis, foliolo terminali oblongo-obovato* acuminato, paniculæ longæ foliosæ ramis brevibus racemosis ascendentibus sed summis et ultraaxillaribus patentibus vel divaricatis, rachi undulato, aculeis e basi longa declinatis validis sed summis tenuibus, *sepalis* lanceolato-attenuatis *fructui laxe adpressis.*

R. Hystrix Weihe in Bluff et Fingerh. Compend. Fl. Germ. i. 687 (1837). Rubi Germ. 92. t. 41. Bab.! Man. ed. 3. 99; ed. 6. 112.

R. Radula β Hystrix Bell Salt.! in A. N. H. xvi. 369; in Fl. Vect. 158.

R. Radula Lindl.! Syn. ed. 2. 94 (in part). Leight.! Fl. Shrop. 232 (in part).

R. Lingua Bab.! Syn. 24; Man. ed. 2. 103.

R. carpinifolius Johnst.! E. Bor. 67.

R. approximatus Quest.! in Billot, Fl. Gall. et Germ. exsic. No. 2454 (sp.).

R. glandulosus γ rosaceus Leight.! Fasc. No. 23 (sp.).

R. pallidus Lees! Malv. 52.

Stem arcuate-prostate angular, rather furrowed throughout, with many nearly equal and short aciculi and setæ, few hairs except near the base. *Prickles* rather strong but slender, declining from very long compressed bases, unequal, chiefly on the angles. *Leaves* quinate-pedate. *Leaflets* flat, wavy at the edge, coarsely and more or less doubly dentate, all oblong-obovate acuminate, pilose above, hairy and a little paler but not felted beneath; midribs and furrowed petioles with many unequal hooked prickles beneath; stipules linear-lanceolate.

Flowering shoot from brown silky scales, setose, aciculate, hairy. *Leaves* ternate. *Leaflets* oblong-obovate, acuminate, pilose above, hairy on the veins beneath, finely but irregularly dentate ; basal nearly sessile, unequal-based; uppermost floral often simple. *Prickles* rather strong, long-based, declining, those of the panicle slender. *Panicle* very long ; branches distant, racemose, mostly axillary; top racemose-corymbose ; rachis and peduncles with many short rather unequal aciculi and setæ and a thin coat of felt near the top. *Sepals* ovate-attenuate, often leaf-pointed, felted, setose, aciculate. *Petals* distant, lanceolate, rounded at the end, entire, pink. *Filaments* white. *Anthers* greenish. *Styles* greenish, but pink at the base. *Primordial fruit-stalk* rather long ; terminal stalks of the branches usually shorter than the lateral stalks.

Our plant differs in some respects from that figured with the same name in the *Rubi Germanici*. The prickles on the stem of the latter are very different, being broad, flat, and narrowing gradually from their base to their tip: there are also many smaller but similar prickles upon the faces of the stem : also the stamens are described as reddish and the sepals as reflexed from the fruit.

A plant received from Mr Watson, others named *R. fuscus* by Lees and Bloxam, and one gathered near Coalford, Gloucestershire, by Mr Hort, have a very thin coat of felt beneath their leaves, and much resemble a specimen sent by M. Questier as *R. Lingua;* but they have a clasping calyx consisting of lanceolate-attenuate (not oval-cuspidate) and much more prickly sepals. They do not associate well with *R. Hystrix*, and will probably be separated from it. I cannot discover a foreign description which agrees with them.

The *R. carpinifolius* of Johnston's *Eastern Borders* is a form of this species, in which the stems are rounder and less hairy, and the short aciculi and setæ are much fewer in number and more deciduous. The stems are "glabrous, but roughish to the touch from obscure points and a few imperfect setæ." But the points are not quite so obscure as he thought; on the older stems they are far from unfrequent and of rather a large size, although but little elevated. The examination of his specimens leaves no doubt upon my mind about the true name of the plant. The misnaming of this plant is one of the results of the unfortunate reference of setose and aciculate plants to *R. carpinifolius*, and especially of the publication of a form of *Koehleri* with the name in Leighton's *Shropshire Rubi*.

My *R. Lingua* from Oakhampton differs very slightly from typical *R. Hystrix*. Its terminal leaflet is more abrupt, and the edges of all the leaflets are much more finely dentate with the principal teeth patent, but the others directed very decidedly towards the apex of the leaflet. Its panicle is very loose, consisting almost wholly of long simple peduncles. I have only seen it when the panicles were very young.

Dr Bell Salter's plant from Parkstone and my *var. tomentosus* belong to *R. Radula*. Neither of these plants has any relationship to *R. scaber*, as seems to have been supposed by Dr Bell Salter, and was suspected by myself.

I place the *R. approximatus* (Quest.) of Billot's *Fl. Gall. et Germ. exsic.* No. 2454 under *R. Hystrix* with scarcely any doubt. Its leaves are broader and rounder than on any British specimen which I have seen. Its panicle also is more open with more ultra-axillary and corymbose branches.

The *R. Hystrix* found at Killarney approaches nearly to it in both of these respects. M. Questier considers his plant as perhaps the same as *R. rudis* ε *denticulatus* (Bab.), which I now believe to be a variety of *R. Radula.* I cannot agree with him in this identification. *R. infestus* (Billot, l. c. No. 2453) closely resembles *R. approximatus* at first sight; but it has none of the characters of the *Radulæ.* I place it with *R. villicaulis.* Indeed it closely approaches the *R. pampinosus* (Lees).

Habitat.—Hedges and thickets. July, August.

Area.—1 2 3 4 5 . . 8 . 10 11 . . . . 19.

Localities.—i. Oakhampton, *N. Dev.;* Tamerton Foliott, *S. Dev.* (Briggs!); Swan Pool, Falmouth, *W. Corn.*—ii. Quarr Wood and St John's, *Isle of Wight.*—iii. Leith Hill and Hook, *Surr.;* Little Berkhampstead, *Herts;* Woodmancote, *W. Suss.;* Harrow, *Middl.* (Hind!).—iv. Cromer, *E. Norf.*—v. Coleford and Stapleton near Bristol, *W. Glouc.;* Wentwood, *Monm.;* Penyard Park wood and Lyston, *Heref.;* Atherstone, *Warw.;* Redwood near Cheltenham, *E. Glouc.* (Notcutt!).—viii. Little Orton and Twycross, *Leic.* —x. Thirsk, *N. E. York.* (Baker!); Aisenby, *N. W. York.*— xi. Ancroft, *Chev.* (Johnst. !).

xix. Killarney, *S. Kerry.*

25. R. rosaceus Weihe.

R. caule arcuato-prostrato angulato, *aculeis*, e basi dilatato-compressa .tenuibus declinatis subæqualibus *nonnullis brevioribus aciculos setas pilosque* inter se subæquales *paulo excedentibus,* foliis quinato-pedatis vel ternatis, foliolis duplicato-dentato-serratis supra pilosis subtus pallidioribus in venis tantum pilosis, foliolo terminali obovato vel oblongo-acuminato basi sæpe subcordato, *paniculæ subpyramidalis* truncatæ infernè foliosæ ramis racemosis ascendentibus sed ultra-axillaribus patentibus corymbosis vel simplicibus, rachi plus minusve undulato, aculeis tenuibus declinatis, sepalis lanceolato-attenuatis fructui laxe adpressis.

R. rosaceus Weihe in Bluff et Fingerh. Compend. Fl. Germ. 685 (1825). Rubi Germ. 85. t. 36. Bor. Fl. Cent. ed. 3. ii. 192. Bell Salt. in Phytol. ii. 133 (the Selborne plant). Bab.! Man. ed. 6. 113.

R. glandulosus γ *rosaceus* Bab.! Man. ed. 2. 105; ed. 4. 105.

R. Güntheri Quest.! in Billot, Fl. Gall. et Germ. exsic. No. 2057 (sp.).

R. glandulosus β *Lejeunii* Bell Salt.! in Brom. Fl. Vect. 159.

R. affinis Sm. Eng. Fl. ii. 405 (teste Borrer!).

R. Koehleri ε *fuscus* Blox.! Fasc. (sp.).

Stem angular, striate, arcuate-prostrate (usually of a bright coral-red colour); hairs, setæ and aciculi rather few, short, nearly equal in length, the two latter seated upon

minute tubercles. *Prickles* all small and slender, unequal,
with long compressed bases, the smaller scattered, the larger
chiefly confined to the angles of the stem. *Leaves* quinate-
pedate or ternate. *Leaflets* doubly and rather coarsely
dentate-serrate, dark green and pilose above, paler and
pilose or hairy on the veins beneath ; basal leaflets shortly
stalked, oblong, bluntish ; intermediate obovate-acuminate ;
terminal broadly obovate or oblong, usually subcordate at
the base, often narrowing uniformly from about the middle
to the end : ternate leaves have the lateral leaflets very
unequal-sided or strongly lobed externally (in the *Rubi
Germanici* the plant is represented with a ternate leaf of
the latter kind) ; petioles flat above ; stipules slender.

Flowering shoot from fuscous ashy scales, rather hairy,
very setose ; prickles slender, declining, from a long compressed base ; aciculi, setæ and hairs many and unequal.
Leaves ternate. *Leaflets* ovate-lanceolate, slightly pilose
above, paler and slightly pilose on the veins beneath ;
lateral shortly stalked. *Panicle* rather pyramidal, blunt or
truncate ; axillary branches several, rather distant, the
uppermost corymbose, the others becoming longer succes-
sively downwards and racemose, all falling short of the
leaves ; ultra-axillary part short and broad, with short
about 3-flowered corymbose or simple and 1-flowered patent
branches ; on the branches the terminal are shorter than
the lateral flower-stalks ; rachis slightly wavy ; peduncles
and branches with slender aciculi, many unequal setæ, a
thin coat of rather adpressed hairs, but no felt. (Sometimes the hairs are more abundant and cross each other so
as greatly to resemble felt, but the fine coat forming real
felt (or tomentum) seems to be always wanting.) *Sepals*
lanceolate-attenuate, green externally with a whitish border,
setose, felted, often slightly aciculate, patent or loosely adpressed to the fruit. *Petals* "pale pink," oblong. *Styles*

purple (?). *Nut* very roundly obovate; inner edge nearly straight.

This plant is very closely allied to *R. Hystrix*; far more so than to *R. glandulosus*, with which it has always been placed by English authors. Its stem is like that of the other *Radulæ*, but the larger prickles are not quite so markedly separated from the small ones as is usually the case. That difference in the stem, and the more elegant, more pyramidal, and more abrupt panicle, are the chief distinctions between it and *R. Hystrix*, with which I strongly suspect that it will prove to be specifically identical. It is with some diffidence that this opinion is given, for the authors who have written concerning the plants seem agreed in placing a considerable interval between them. Mr Notcutt, after studying the two plants in Red Wood, near Cheltenham, was of opinion that they form only one species.

R. Radula has a similar stem, but its prickles are not so unequal; its leaves are finely toothed and felted beneath; its panicle has not the pyramidal form, nor its sepals the leaflike point.

We learn from a specimen in Borrer's Herbarium (gathered at Woodmancote, Sussex) that he referred this plant to *R. Koehleri γ pallidus*, and believed it to be the *R. affinis* of Smith. The panicle of that specimen is very much divided and exceedingly prickly. Mr Edw. Forster considered the *R. Koehleri α* to be the plant intended by Smith. I have already expressed my belief (p. 75) that Smith's *R. affinis* is our *R. Lindleianus*.

M. Questier sends this plant under the name of *R. Güntheri*, from which it differs in many respects : for instance, in the armature of its stem, the form of its panicle, and the total want of felt beneath its leaves.

Habitat.—Woods and hedges. July, August.

Area.—1 2 3 . 5 . 7 8 . 10 11 12 . . . . . . 19 . . . . . 25.

Localities.—i. Linton and South Molton, *N. Devon* (Bell Salt.!).—ii. Guildford, *Isle of Wight;* West Chiltington, *E. Suss.;* Woodmancote, *W. Suss.;* The Lyth, Selborne, *S. Hants.*—iii. Easney Park wood, *Herts;* Pinner wood and Harrow, *Middl.* (Hind!); Thames Ditton, *Surr.*—v. By the Buckstone near Monmouth, *W. Glouc.;* Red wood near Cheltenham, *E. Glouc.;* Chase wood near Ross, *Heref.* (Purchas!).—vii. Bangor and by Pen Maen Mawr, *Caern.* —viii. Twycross, *Leic.*—x. Terrington Car and Thirsk, *N. E. York.;* Loxley near Sheffield, *S. W. York.*—xi. Holliwell Dene, *Northumb.*—xii. Rydal falls, *Westm.*

xix. Killarney, *S. Kerry.*—xxv. Ladiston, *W. Meath* (D. Moore).

Jersey.

26. R. pygmæus Weihe.

R. caule arcuato-prostrato subtereti, *aculeis* crebris tenuibus inæqualibus declinatis *basi paululum dilatatis*, aciculis tenuissimis et sétis pilisque inæqualibus crebris, foliis quinato-pedatis vel ternatis, *foliolis grossè et inæqualiter duplicato-serratis* supra pilosis *subtus* pallidioribus *in venis tantum pilosis*, foliolo terminali obovato-acuminato, paniculæ coarctatæ infernè foliosæ ramis corymbosis, rachide recta, aculeis tenuibus declinatis, aciculis setis pilisque inæqualibus crebris, *sepalis ovato-attenuatis aciculatis longi-setosis* tomentosis a fructu laxe reflexis.

R. pygmæus Weihe in Bluff et Fingerh. Compend. Fl. Germ. 687 (1825). Rubi Germ. 93. t. 42. Wimm. et Grab. Fl. Siles. ii. 43. Bab.! Man. ed. 6. 113.

R. hirtus β *Menkii* Bab.! Syn. 29 ; Ann. Nat. Hist. Ser. 2. ii. 39 ; Man. ed. 2. 105 ; ed. 4. 105.

Stem nearly terete, prostrate (?). *Prickles* many, very unequal, very slender, declining, flattened but only slightly dilated at the purplish base, otherwise yellow. *Aciculi* very slender, and as well as the *setæ* and *hairs* many, unequal, patent ; or the hairs slightly crisped. *Leaves* quinate-pedate or ternate. *Leaflets* coarsely irregularly or somewhat doubly serrate, green and pilose above, slightly paler and pilose on the veins but not felted beneath ; basal lanceolate ; intermediate broadly lanceolate, attenuate ; terminal obovate-

16

acuminate-attenuate, subcordate below; all stalked : all
nearly equal on the ternate leaves; lateral unequal-sided
or slightly lobed, ascending ; midribs and petioles with
small slender slightly hooked prickles beneath ; stipules
very slender, linear.

Flowering shoot from brown ashy scales, armed like the
stem but less strongly. *Leaves* ternate. *Leaflets* obovate-
lanceolate-acuminate, dull and pilose above, very slightly
paler and hairy on the veins beneath ; few uppermost floral
leaves simple. *Panicle* rather long ; axillary branches distant, short, corymbose, erect; top short, leafless, close, racemose, with very short corymbose often divaricate branches ;
terminal flowers of branches and panicle with longish stalks ;
rachis and branches and pedicels hairy, scarcely if at all
felted, with many prominent very unequal setæ and very
slender nearly straight unequal subpatent prickles. *Sepals*
ovate-attenuate with a slender point, greenish, felted, hairy,
with many prominent setæ and aciculi, loosely reflexed from
the fruit. *Petals* oval or lanceolate, narrowed below, entire,
white or pinkish. *Filaments* white. *Anthers* purple. *Styles*
greenish. *Primordial fruit-stalk* shorter than the calyx.

I have long had much doubt concerning the true position
of this plant, and am now surprised that it should have been
considered as *R. Menkii* (Weihe) ; for a careful comparison
of the plate and description given in the *Rubi Germanici*,
with specimens received from Mr Borrer and the Rev.
W. M. Hind, convinces me that they are not even very
nearly related. These specimens accord so well with *R.
pygmæus* (Weihe) as described and figured in the *Rubi
Germanici*, that there seems very little reason to doubt the
specific identity of the plants. Should this plant prove
abundant near Tonbridge or Watford, it will probably be
rightly considered as a species distinct from our other brambles. It seems more nearly related to *R. rosaceus* than to

any other of our plants. A specimen of what is perhaps the true *R. Menkii* will be found in Billot's *Fl. Gall. et Germ. exsic.* No. 1868.

Habitat.—Hedges and woods. July, August.

Area.— . . 3.

Localities.—Mount Nod and Eridge near Tonbridge Wells, *Kent;* Oxhey Wood, Watford, *Herts;* Pinner Wood, *Middl.*

27. R. scaber Weihe.

R. caule arcuato-prostrato subangulato subsulcato, *aculeis validis brevibus* subæqualibus e basi longa compressa *declinatis deflexisve*, aculeis setis pilisque paucis brevissimis, foliis ternatis vel quinatis, foliolis duplicato-dentatis supra opacis pilosis subtus pallide viridibus pilosis, foliolo terminali late obovato cuspidato vel acuminato basi subcordato, paniculæ subpyramidatæ foliosæ tomentosæ apice truncato vel obtuso ramis ultra-axillaribus racemoso-corymbosis vel simplicibus axillaribus erecto-patentibus racemosis, aculeis brevibus e basi longa declinatis vel deflexis, aciculis validis, setis pilisque subæqualibus, sepalis ovato-acuminatis a fructu laxe reflexis.

R. scaber Weihe in Bluff et Fingerh. Compend. Fl. Germ. 683 (1825). Rubi Germ. 80. t. 32. Bab.! in A. N. H. ser. 2. ii. 41; Man. ed. 3. 103; ed. 6. 113. Blox.! in Kirby, 41. Bor. Fl. Centre, 190.

R. Babingtonii Bell Salt.! in A. N. H. xv. 307 (1845); Phytol. ii. 138. Bab.! Syn. 21; Man. ed. 2. 102; ed. 3. 99. Blox.! in Kirby, 40.

R. Kaltenbachii Metsch in Linnæa, xxviii. 170 (1856). Wirtg.! Herb. Rub. No. 92 (sp.).

R. Löhrii Wirtg.! Fl. Preuss. Rhein. 162; Herb. Rub. No. 22 (sp.)?

R. mutabilis Genev.! Essai, 5 (1860).

Stem arcuate-prostrate, round at the base, angular above with the angles rounded and the faces often sulcate. *Prickles* short, slightly deflexed or much declining, from a compressed base, of which the longer diameter often equals or even exceeds the length of the prickle, rather many. *Aciculi, setæ* and *hairs* rather few, very short, with large prominent bases, which give a coarse file-like roughness to the old stems. *Leaves* ternate or quinate. *Leaflets* all stalked, slightly concave, cuspidate, rather doubly dentate (that is, about every fourth tooth is larger than the rest and patent, or even turned from the tip of the leaf towards which the others slightly incline; when the toothing is very coarse this arrangement is often much less or scarcely at all apparent; but I have seen a specimen in which the large teeth have so much increased in size as to absorb the other teeth and render the leaves doubly and patently dentate or nearly lobate-dentate; the tip of each of these large double teeth being hooked backwards), opaque, deep green and pilose above, paler and hairy on the veins beneath, usually very large; basal oblong; intermediate obovate; terminal roundly obovate : the lateral leaflets of the ternate leaves are almost as large as the terminal leaflet and very strongly lobed on their outer side; petioles which are flattened above, and midribs with many hooked prickles beneath; stipules slender, linear-lanceolate.

Flowering shoot from silvery scales, hairy, setose, prickly like the stem but less strongly. *Leaves* ternate. *Leaflets* resembling those of the stem ; basal ovate, unequal-sided; terminal ovate-acuminate, narrowed below; or all subcordate below; uppermost floral leaves simple, cordate-ovate. *Panicle* long, very hairy and setose and with ash-coloured felt towards the top; prickles small, usually strong, thick-based, declining, the uppermost more slender ; axillary branches racemose, ascending, few-flowered ; ultra-axillary few, corymbose, few-flowered, or 1-flowered, patent, usually forming a rather close ovate top. *Sepals* lanceolate-attenuate, with a narrow leaflike point, hairy, aciculate, setose, felted, greenish, loosely reflexed from the fruit. *Petals* oblong, rather acute, narrowed below, white. *Filaments* white. *Anthers* and *styles* greenish. *Primordial fruit-stalk* shorter than the sepals, other peduncles often long. Fruit large, ovate, well-flavoured. *Nut* ½-oval; inner edge nearly straight.

In the smaller forms (*R. scaber* Bab.) the panicle is often nearly or quite simple in its upper part, the peduncles which spring directly from the rachis being 1—1½ inches long and divaricate. Rarely the branches gradually lengthen downwards, and give a somewhat pyramidal form to the panicle. Usually, in large as well as small states of the plant, the panicle is narrow throughout. In the larger states (*R. Babingtonii* Bell Salt.) the ultra-axillary part is much less in proportion to the whole of the very long panicle, and the simple peduncles are much fewer and shorter. These are often enormous plants with very long prostrate exceedingly rough stems, a panicle not unfrequently more than three feet long with the lower branches forming secondary panicles, and large floral leaves.

The *R. Löhrii* (Wirtg.), which is shortly described and illustrated by a specimen in the valuable *Herb. Ruborum rhenanarum* (No. 22), is very closely allied to *R. scaber;* much approaching what was once called *R. Babingtonii.* The under side of its leaves is totally devoid of felt, as is the case in our plant, although the very dense hairs seated even upon the smaller veins not unfrequently give to it an appearance of being felted. *R. Löhrii* has the branches of the panicle more decidedly corymbose, with the exception of the very lowest, than is usual on British specimens, although the panicle of some approaches very closely to that structure. In all these respects I see no material difficulty attending the combination of *R. Löhrii* with *R. scaber;* but there remains the fact that Wirtgen finds the sepals of his *R. Löhrii* to be "fructus erectis;" and it is for other botanists to judge if this is a sufficient reason for retaining it as a distinct species: I suspect not.

A careful study of the very full account of *R. Kaltenbachii* given in the *Linnæa* convinces me that it is exactly synonymous with *R. Babingtonii* (Bell Salt.), and that the true *R. scaber* (Weihe) differs more from it than I had supposed. Without the advantage of examining authentic *R. scaber* it is impossible to be certain concerning the specific identity of it and *R. Babingtonii*, although my opinion tends strongly to the belief that they form the extreme states of one species. Such is not the opinion of Dr Metsch, who possessed better opportunities of becoming acquainted with the *R. scaber* (Weihe). He thinks his *R. Kaltenbachii* (*R. Babingtonii*) quite distinct from *R. scaber*. The authentic specimen in my copy of Wirtgen's *Herb. Rub.* has an imperfectly developed panicle. Should the opinion of Dr Metsch be correct the plants will have to bear the name given by Bell Salter in 1845, eleven years before the publication of the *R. Kaltenbachii*. But when the characters of even these extreme states, as I think them, of *R. scaber* are carefully contrasted, it will be found that there is very little real difference between them, and a well chosen series of specimens leaves little doubt concerning the specific identity of our *R. scaber* and the *R. Babingtonii*, and judging from the description, the *R. Kaltenbachii*. It is possible that our small plant is not really the *R. scaber* of Germany, although it appears to agree very well with the description and plate in the *Rubi Germanici*.

The intermediate forms agree well with the *R. mutabilis* (Genev.!), although the authentic specimens of that plant have more felt on the leaves than is usual in our forms of *R. scaber*, especially on those of the flowering shoot. I have seen traces of felt on some of the English specimens referred to *R. scaber*, but never so much as is found on the *R. mutabilis*, from Cleves, North Yorkshire, which was so named by M. Genevier. But another specimen gathered at the same place in the succeeding year has no felt on its leaves, and is referred to *R. scaber* with much confidence.

I am not acquainted with the *var. β verrucosus* of Lees (*Bot. Worc.* 43), which he says grows "in profusion on Bromsgrove Lickey.... mixed with no other bramble." The only specimen which I have from that place is named by Mr Lees *R. affinis β effusus*. It does not agree with the description of the *var. verrucosus*, and is, I believe, a form of *R. Sprengelii (R. Borreri).*

It is not likely that much objection will be raised to the position in the genus which is now assigned to *R. scaber*. The *R. Babingtonii* was usually placed amongst the *Radulæ*, and the small *R. scaber* never seemed to agree with the *Bellardiani.*

Several plants have been erroneously called *R. Babingtonii:* that from Great Cowley Park is *R. amplificatus;* from Monmouth, is *R. Colemanni,* noticed above under *R. Grabowskii;* that from Shrawley wood, Worcestershire (*R. longithyrsiger,* Lees MS.), is *R. pyramidalis.*

Habitat.—Woods. July, August.

Area.—1 2 3 . 5 . 7 8 . 10.

Localities.—i. Leigh woods near Bristol, *N. Som.* (T. B. Flower).—ii. Selborne, *S. Hants.;* Henfield, *W. Suss.* (Borr!). —iii. Essendon, Easney Park wood, and Praë wood, *Herts;* Horsenton wood near Harrow, *Middl.;* Rivenhall, *N. Essex* (Varenne!).—v. Chepstow, *Monm.;* Cirencester road near Cheltenham, *E. Glouc.* (Notcutt); Hartshill wood, *Warw.;* Wire Forest, *Worc.;* Seckley wood and Shawbury Heath, *Salop.*—vii. Llanberis, *Caern.*—viii. Charnwood, and Buddon wood (Bloxam), *Leic.;* Calke, *Derb.*—x. Cleves, *N. E. York.*

28. R. rudis Weihe.

R. caule arcuato angulato subsulcato, aculeis validis conico-compressis subæqualibus subpatentibus *aciculos setas pilosque inter se subæquales* et breves excedentibus, foliis quinatis, *foliolis grosse et duplicato serratis* (vel lobato-serratis) *subtus viridi-albo-tomentosis,* foliolo terminali elliptico vel late oblongo-obovato acuminato, paniculæ longæ foliosæ ramis ascendentibus corymboso-racemosis, ramis summis et ultra-axillaribus divaricatis, rachide recto, aculeis è basi longa declinato-deflexis validis summis tenuibus, sepalis ovato-attenuatis a fructu reflexis.

R. rudis Weihe in Bluff et Fingerh. Compend. Fl. Germ. i. 687 (1825). Rubi Germ. 91. t. 40. Lindl.! Syn. ed. 1. 94. Bab.! Man. ed. 3. 100; ed. 6. 114. Blox.! in Kirby, 41. Bell Salt.! in Fl. Vect. 158 (excl. var. δ). Lees! Malv. 54. Metsch in Linnæa, xxviii. 196. Leight. Shrop. Rubi, 16 (sp.).

R. rudis ε *attenuatus* Bab.! Man. ed. 2. 102.

R. rudis δ *Reichenbachii* Bab.! Man. ed. 4. 103; Syn. 22.

R. echinatus Bab.! Man. ed. 1. 96. Lindl.! Syn. ed. 1. 94; ed. 2. 93. Borr.! in Hook. Br. Fl. ed. 2. 247; ed. 3. 251.

R. Radula Leight.! Fl. Shrop. 232 (in part).

R. Radula γ *Hystrix* Bab.! Man. ed. 1. 96.

R. Radula γ *pygmæus* Bab.! Syn. 24; Man. ed. 2. 103.

R. discerptus Müll.! Mon. 73. Chab. Étude du Rub. 26.

R. Genevierii Bor.! Fl. du Centre, ed. 3. t. 193.

R. bracteatus Billot! Fl. Gall. et Germ. No. 1470 (sp.).

Stem arcuate, rough, angular, furrowed. *Prickles* nearly equal, slender but strong, compressed-conical, from a slightly enlarged base, subpatent, chiefly placed on the angles of the stem and scarcely passing into aciculi. *Aciculi, setæ,* and *hairs* many, very short, nearly equal. *Leaves* quinate. *Leaflets* rather concave, deeply and coarsely dentate-serrate or almost lobed, hairy, and having a dull lurid appearance above, pale with many hairs on the ribs and much greenish-white felt beneath; basal shortly-stalked, obovate-lanceolate; intermediate stalked, obovate-lanceolate acuminate; terminal oval or elliptic-obovate, acuminate; petioles which are flat above, and midribs with short hooked prickles beneath; stipules linear-lanceolate.

Flowering shoot from fuscous scales, slightly wavy, hairy, setose, aciculate. *Prickles* strong, from large bases, declining or slightly deflexed. *Leaves* mostly ternate. *Leaflets* much narrowed below, acuminate, greenish-white and felted beneath. *Panicle* narrow, leafy, aciculate, setose; upper prickles slender; branches short, ascending, between racemose and corymbose, few-flowered, mostly axillary; top corymbose. *Sepals* ovate-attenuate, aciculate, setose, hairy, felted, often leaf-pointed. *Petals* distant, oblong, white. *Filaments* white. *Anthers* and *styles* greenish. *Primordial fruit-stalk* about as long as or slightly longer than the reflexed sepals. *Nut* ½-obovate; inner edge nearly straight.

The sulcate stem, nearly equal and scarcely scattered prickles, short and nearly equal aciculi, setæ, and hairs, coarsely serrate or even jagged leaflets, and strongly reflexed and usually leaf-pointed sepals, distinguish this bramble from *R. Radula.* It is separated from *R. .Hystrix* by the felted under side of its leaves and the sepals being so much reflexed from the fruit as usually to press closely against the peduncle. Rarely, especially when it grows in deep shade, the felt on the leaves seems to be wanting, and

then it is difficult to draw up characters by which to distinguish it with certainty; nevertheless the practised eye can
hardly be deceived even in such cases. An example of this
naked state is given as *R. rudis forma umbrosa* in *Wirtg.
Herb. Rub.* No. 90.

The plant from Bangor which I formerly called *R.
Reichenbachii* was incorrectly so named; for that of Weihe
has a stem which is "aciculis et glandulis destitutus." My
plant has few of them, but they are far from being wanting.
I still think that it is a form of *R. rudis*, with rounder leaflets and a broader and more compound panicle than is found
in the typical plant.

The *R. Radula* γ *pygmæus* of my *Synopsis* seems to be a
state of *R. rudis* in which the toothing of the leaves is very
much reduced in size; but the specimen from Bristol which
was so named is *Koehleri* γ *pallidus*. *R. Leightonii* and my
variety named *denticulatus* are now placed under *R. Radula*,
to which species they seem to be much more closely allied
than to *R. rudis*.

The variety *microphyllus* mentioned by Bloxam (*Kirby's
Flora*, 41) is an elegant state of *R. rudis* with remarkably
small leaves, which are serrated similarly to those of the
typical plant, but much more finely.

The plant named *R. Radula* by Nees for Leighton is
certainly my *R. rudis* in the state called *R. echinatus* (*Man.*
ed. 1), and the *R. echinatus* of Lindley in the second edition
of his *Synopsis*. It is also the *R. echinatus* mentioned by
Borrer. The specimens from the Horticultural Society's
Garden of *R. rudis* and of *R. echinatus*, derived from Lindley's authentic bushes (*Herb. Borr.*), are identical, and are
most certainly both *R. rudis*.

There is a specimen in *Herb. Borr.*, from Ninham in the
Isle of Wight, gathered by Dr Bell Salter, and named by
him "probably *R. Radula*," which appears to belong to *R.*

rudis.   It agrees in most respects with another, from Poole, Dorset, also called *R. Radula* by him.

Mr Baker has sent specimens with the name of *R. rudis β Leightonii,* gathered at Aislaby and Sowerby near Thirsk in Yorkshire, which seem to be correctly referred to this species.   Their leaves differ considerably from the typical form ; the terminal leaflet is roundly obovate-acuminate with a subcordate base ; the toothing is less coarse and less deep ; the veins are only slightly hairy beneath, although the whole under surface is finely felted.   The leaves of the flowering shoot have a very thin coat of felt.   The panicle is not so narrow as that of the typical plant.

I know only the old leafless stem and panicle of Dr Moore's plant from the county of Wicklow, but am tolerably certain of its being *R. rudis.*

I possess a specimen gathered at Kullaberg, Scania, Sweden, in 1846, by A. G. Longberg, and named *R. Radula.*   It appears to be *R. rudis,* a species not recorded as Swedish by Arrhenius or Fries.

I place the *R. bracteatus* (Billot ! *Fl. Gal. et Germ. exsic.* No. 1470) with *R. rudis.*   The only specimen that I have seen is very like a form not unfrequently assumed by our plant.   M. Boreau is of the same opinion (*Billot, Annot.* 93).

In common with Mr J. G. Baker I am unable to distinguish the *R. discerptus* (Müll.) from *R. rudis,* nor can the *R. Genevierii* (Bor. !) be separated from this species.

Habitat.—Hedges and thickets.   July, August.

Area.—1 2 3 . 5 6 7 8 . 10 11 12 13 14 . . . . 22 23.

Localities.—i. Plymouth, *S. Dev.* (Briggs); Leigh woods near Bristol, *N. Som.*—ii. *Isle of Wight* (Borr. ! and Bell Salt. !);   Poole, *Dors.* (Bell Salt. !) ;   Bere Forest, *Hants* (Hind !);   Great Ridge near Boyton, *S. Wilts* (E. Forster !). —iii. Long Ditton and Chertsey, *Surr.;*   Goldings, *Herts. ;*

Trent Park, *Middl.* (Hind !).—v. Lanwarne and Ross,
Heref.; Cowleigh Park, *Worc.;* Almond Park, Haughmond, and Berwick, *Salop;* Lydney, *W. Glouc.*—vi. Cardigan, *Card.*—vii. Welshpool, *Montg.;* Banks of the Menai
near Bangor, *Caern.*—viii. Twycross, *Leic.*—x. Thirsk, *N. E.
York.*—xi. Howick, *Northum.* (Borr.).—xii. Rydal, *Westm.*
(Dr Cookson).

xiii.—Gourock, *Renf.*—Repton, *Berw.*

xxii.— *Wicklow* (D. Moore !).—xxiii. New Grange, *Meath.*

29. R. Radula Weihe.

R. caule arcuato angulato, aculeis e basi dilatato-
compressa tenuibus declinatis *aciculos pilos setasque*
multos breves sed *inter se inæquales* excedentibus, foliis
quinato-pedatis, *foliolis argute* sed duplicato-patenti-
dentatis subtus viridi-albo-tomentosis, foliolo terminali
obovato-acuminato vel subcuspidato, paniculæ longæ
foliosæ ramis brevibus corymbosis ascendentibus, aculeis
e basi longa declinatis validis summis tenuibus, sepalis
ovatis a fructu laxe reflexis.

a verus; aculeis caulium sterilium inæqualibus,
foliolo terminali obovato-acuminato.

R. *Radula* Weihe in Bluff et Fingerh. Compend. Fl.
Germ. i. 686 (1825).   Rubi Germ. 89. t. 39.   Arrh.! in
Fr. Summa, 116.   Fries! Herb. Norm. viii. 47 (sp.).   Sond.
Fl. Hamb. 280.   Bab.! Man. ed. 3. 99; ed. 6. 114.   Bell
Salt. in Fl. Vect. 158 (excl. β).   Blox! in Kirby, 42.   Bor.
Fl. Centre, 191.   Metsch in Linnæa, xxviii. 190.   Wirtg.!
Herb. Rub. 88 (sp.).   Syme, E. B. iii. 184. t. 452.   Merc.
in Reut. Cat. Pl. Genev. 273.
R. *villicaulis* ε *pubescens* Bab.! Man. ed. 1. 95.
R. *fusco-ater* β *candicans* Bab.! Syn. 25; Man. ed. 2. 104.

β *Leightonii;* aculeis caulium sterilium subæqualibus, foliolo terminali obovato-cuspidato.

R. *Leightonii* Lees! in Leight. Fl. Shrop. 233 (1841).
Bab.! Man. ed. 1. 96.

R. rudis β *Leightonii* Bell Salt.! in A. N. H. xvi. 367.
Bab.! Syn. 22; Man. ed. 4. 102. Leight.! Shrop. Rubi
17.(sp.).

R. Lingua β *tomentosus* Bab.! Syn. 25; Man. ed. 2. 103.

R. Radula v. *sylvaticus* Wirtg.! Herb. Rub. No. 89 (sp.).

R. Radula β *Leightonii* Bab. Man. ed. 5. 105; ed. 6. 114.

γ *denticulatus;* foliolo terminali late quadrangulari-
obovato cuspidato basi subcordato late sed inepte den-
tato dentibus denticulatis.

R. Radula γ *denticulatus* Bab.! Man. ed. 5. 105; Man.
ed. 6. 114.

R. rudis ε *denticulatus* Bab.! in A. N. H. xix. 87.

Stem arcuate, round at the base, angular above. *Prickles*
unequal, large, conical, patent, from a large compressed base.
Aciculi and *setæ* many; *hairs* rather fewer; all short but
unequal. *Leaves* quinate-pedate. *Leaflets* slightly convex,
dull and glabrous above, whitish and hairy and felted beneath, doubly but finely dentate-serrate with the primary
teeth usually subpatent, wavy at the edge; basal lanceolate;
intermediate obovate, acuminate; terminal roundly obovate,
acuminate-cuspidate; midribs and petioles with many, unequal, hooked prickles beneath; stipules linear-lanceolate;
petioles flat above.

Flowering shoot from fuscous scales, slightly wavy, hairy,
slightly setose and aciculate. *Prickles* strong, from a large
base, declining. *Leaves* quinate or ternate. *Leaflets* usually
clothed like those of the stem, but sometimes nearly without felt; uppermost floral leaves often simple, broadly cor-
date-3-lobed-acuminate or ovate-acuminate. *Panicle* narrow,
leafy, very hairy, aciculate, setose, felted; prickles of rachis
slender; branches short, corymbose, ascending, axillary; top
racemose, with few short, patent, few-flowered, corymbose,

ascending branches. *Sepals* ovate-acuminate, aciculate, setose, hairy, felted, with a slender point. *Petals* distinct, oblong, narrowed below, notched at the blunt round end, pink. *Filaments* pale pink. *Anthers* greenish yellow. *Styles* pink. *Primordial fruit-stalk* short, shorter than the loosely reflexed sepals. *Nut* ovate ; inner edge nearly straight.

The *R. Leightonii* seems certainly to be a form of this species from which its differences are slight and are chiefly as follows :—Aciculi, setæ and hairs on the stem much fewer, scattered, shorter. Leaflets flat, pilose above ; intermediate cuspidate ; terminal obovate, cuspidate, slightly cordate below. Primordial fruit-stalk shorter than the sepals. It is not constant to these characters. I have never seen any specimens of nearly so marked a kind as those from the two original bushes sent to Lindley and considered as a hybrid by him, unless "really a plant of common occurrence." In my opinion those issued in Leighton's *Fasc. of Rubi* are far from being typical. It is perhaps hardly worthy of separation from *R. Radula* even as a variety. The *R. Radula v. sylvaticus* (Wirtg.), of which a specimen will be found in the *Herb. Rub.* (No. 89), agrees very well with *R. Leightonii*. My *R. Lingua β tomentosus* appears to be *R. Leightonii*. The other variety of my *R. Lingua* will be found noticed under *R. Hystrix*. Mr Bloxam considers the *R. Leightonii* to be the same as *R. melanoxylon* (Müll.), and a distinct species.

The variety called *denticulatus* also varies very slightly from the type of this species. It chiefly differs by the extreme fineness of the felt on its leaflets, which seems indeed to be sometimes altogether wanting ; their very fine dentition, which is nevertheless certainly double; and the very square form of the terminal leaflet. Its prickles are usually yellow, but in one of its states they are of a beautiful blood-red colour. Mr Bloxam's plant, which is noticed in

Syme's *English Botany* (iii. p. 184), does not accord with the true *v. denticulatus*. Some remarks upon this plant will be found under *R. diversifolius*.

Dr Metsch tells us that the stem of *R. Radula* is arcuate-decumbent, but I believe that our plant is truly arcuate, as is that of the *Rubi Germanici*.

Habitat.—Hedges. July, August.

Area.—1 2 3 4 5 6 . 8 . 10 11 . 13 14 . . . .
. . . . 30.

Localities.—i. Pomphleet and other places near Plymouth, *S. Dev.* (Briggs !).—ii. St John's, *Isle of Wight;* Parkstone near Poole, *Dors.;* Henfield, *W. Suss.*—iii. Claygate and Kew Lane, *Surr.;* Harrow, *Middl.* (Hind).—iv. Sandy, *Beds;* Eversden Wood, *Cambs.*—v. Shrewsbury and Shawbury Heath, *Salop.*—vi. St Issels, *Pemb.*—viii. Twycross, *Leic.*—x. Loxley near Sheffield, *S. W. York;* Thirsk, Byland and Bilsdale, *N. E. York.*—xi. Alne, *Northumb.*

xiii. Jardine Hall, *Dumf.*—xiv. Braid Hills and Morton Hall Wood, *Edinb.*

xxx. Kilrea, *Derry* (D. Moore).

Jersey.

GROUP IV. GLANDULOSI.

Caules arcuato-prostrati vel prostrati, radicantes, hirti. Aculei copiosi, valde inæquales, sparsi, in aciculos setasque copiosos graduatim adeuntes.

All the plants included in this section have nearly or quite prostrate stems. They may usually be easily known from those belonging to other sections by their abundant and unequal aciculi and setæ, which graduate into each other and into unequal prickles; also by the want of bloom upon the stem. They may be divided in a tolerably natural manner into the following subordinate groups.

a. *Koehleriani.* Folia quinata vel raro ternata. Aculei, aciculi setæque ad basin incrassati.

b. *Bellardiani.* Folia ternata vel quinato-pedata; foliola infima intermediis dissita, petiolata. Aculei in caulis angulis sæpe congesti. Caules valde hirsuti, aciculati et setosi.

a. *Koehleriani.* Folia quinata vel raro ternata. Aculei, aciculi setæque ad basin incrassati.

The *Rubi Koehleriani* resemble the *Radulæ* in many respects, and are sometimes distinguished from them with difficulty. I am unable to point out any character which will always prove trustworthy. The *Radulæ* rarely, if ever, have the strong aciculiform setæ which connect the true setæ with the aciculi of the *Koehleriani;* neither is it at all usual to find their prickles very unequal in size and length or otherwise placed than on the angles of the stem. Also, the old stems of the *Radulæ* are rough like a file from the presence of abundant low slightly conical tubercles from which the aciculi, setæ and hairs have fallen : in the *Koehleriani* those organs are persistent, but are usually broken from the old stems so as to leave very short blunt prickles in the positions occupied by the tubercles of the *Radulæ.*

30. R. Koehleri Weihe.

R. caule arcuato-prostrato subtereti vel angulato
piloso, aculeis valde inæqualibus e basi compressa
paululum declinatis, aciculis setisque valde inæqualibus,
foliolis inæqualiter vel subduplicato-dentatis *supra
planis subtus* pallide viridibus *in venis pilosis*, foliolo
terminali cordato-ovato *infimis* petiolatis *intermediis
dissitis*, paniculæ apertæ foliosæ ramis brevibus patentibus corymbosis vel ramis axillaribus racemosis, aculeis
crebris longis tenuibus declinatis, aciculis setis pilisque
multis inæqualibus, *sepalis* ovato-attenuatis *patentibus*
vel a fructu reflexis.

R. Koehleri Borr.! in Hook, ed. 2. 247 ; ed. 3. 250.
Bab.! Syn. 26 ; Man. ed. 2. 104 ; ed. 6. 114.   Syme's Eng.
Bot. iii. 185.

a verus; aculeis aciculis setisque multis, *foliolis
subtus asperis in venis tantum pilosis*, paniculæ apertæ
truncatæ sæpe ad apicem dilatatæ ramis longis corymbosis patentibus vel ramis axillaribus racemosis et ascendentibus, *pedunculo terminali paniculæ et ramorum
quam lateralibus breviori.*

R. Koehleri a Bab.! Man. ed. 5. 106 ; ed. 6. 115.
R. Koehleri Weihe in Bluff et Fingerh. Compend. Fl.
Germ. i. 681 (1825).   Rubi German. 71. t. 25.   Borr.! in
E. B. S. t. 2605.   Lindl.! Syn. ed. 1. 94 ; ed. 2. 93.
Metsch in Linnæa, xxviii. 183.   Syme's Eng. Bot. t. 453.

R. echinatus Lindl.! Syn. ed. 1. 94. Leight.! Fl. Shrop. 235.

R. fusco-ater γ *echinatus* Bab.! Syn. 26; Man. ed. 2. 104. Leight.! Shrop. Rubi, 19 (sp.).

R. rudis Lindl.! Syn. ed. 2. 93.

R. glandulosus Sm.! Eng. Fl. ii. 403 (excl. Syn.).

R. Koehleri γ *pallidus* Leight.! Shrop. Rubi 20, (sp.).

R. Koehleri β *cuspidatus* Bab. Syn. 27 ; Man. ed. 2. 104.

R. pallidus Lindl.! Syn. ed. 1. 94. Esenbech! in Leight. Herb. (sp.).

Stem arching rather highly at the base, afterwards prostrate, roundish. *Prickles* many, very unequal, straight, usually slightly declining, from long compressed bases. *Aciculi* strong, resembling small prickles and springing from bases which are similar to but usually shorter than those of the prickles. The stronger *setæ* closely resembling the aciculi, the weaker like the hairs. *Leaves* quinate or very rarely ternate. *Leaflets* convex, dark green and nearly or quite glabrous above, paler beneath, with many short hairs upon the veins, but usually glabrous between them, doubly and often rather patently dentate ; basal directed backwards, oval, acute, stalked ; intermediate obovate, acuminate or slightly cuspidate ; terminal broadly obovate, acuminate or slightly cuspidate, rather cordate at the base ; petioles (which are furrowed above) and midribs with many deflexed prickles beneath; stipules linear-lanceolate.

Flowering shoot from fuscous scales, armed like the stem, but usually with more aciculi, setæ and hairs. *Leaves* ternate. *Leaflets* stalked, obovate, acuminate or obovate-lanceolate, clothed and toothed like those of the stem. *Panicle* long, leafy, truncate, with short ascending usually racemose axillary branches ; ultra-axillary branches rather long, patent, corymbose ; rachis and peduncles usually very prickly and setose. *Sepals* aciculate, setose, ashy, ovate-attenuate, leaf-pointed. *Petals* narrow, distant, obovate, acute, clawed, pink. *Filaments* white. *Anthers* and *styles* pale yellow. *Primordial fruit-stalk* not so long as the sepals; also the terminal fruit-stalk of the branches is usually shorter than the lateral ones, which are commonly divaricate. *Nut* ½-ovate; inner edge nearly straight.

β *infestus;* aculeis aciculis setisque multis validis, *foliolis subtus mollibus in venis tantum hirtis, paniculæ* latæ *ad apicem rotundatæ* ramis mediocribus subcorymbosis erectis sed ramis axillaribus corymbosis erecto-patentibus, *pedunculo terminali paniculæ et ramorum quam lateralibus breviori,* aculeis paniculæ validis deflexis.

R. Koehleri δ *infestus* Bab.! Syn. 27; Man. ed. 2. 104; ed. 5. 106; ed. 6. 115. Lees in Steele, 56.

R. pallidus β *infestus* Bab.! Man. ed. 3. 100; ed. 4. 103.

R. carpinifolius Leight.! Fl. Shrop. 229; Shropshire Rubi! (sp.). (Not of Eng. Bot. Suppl. nor Rubi Germ.)

R. fusco-ater γ *aculeatus* Bab.! Man. ed. 3. 101; ed. 4. 104.

Stem arching rather highly at the base, afterwards prostrate, angular. *Prickles* many, strong, deflexed, or strongly declining, compressed, from very long compressed bases, very unequal. *Aciculi, setæ* and *hairs* moderate in number, short. *Leaves* quinate. *Leaflets* all stalked, dark green and pilose above, hairy on even the smallest veins beneath, and rarely a little felted, flat, wavy at the edge, doubly dentate-serrate, cuspidate; basal oblong; intermediate obovate; terminal shortly obovate; petioles and midribs with many short strong much-hooked prickles beneath; stipules linear lanceolate.

Flowering shoot from fuscous scales, hairy, very prickly, armed like the stem. *Leaves* ternate ; or upper floral leaves simple, cordate-ovate or broadly three-lobed. *Leaflets* obovate, the lateral acuminate, the terminal cuspidate, clothed and toothed like those of the stem. *Panicle* rather long, very prickly, usually with a broad convex top ; branches mostly axillary, short, few-flowered, corymbose. *Sepals* ovate-cuspidate, aciculate, setose, felted, leaf-pointed. *Petals* broadly ovate, clawed, blunt, and slightly notched at the end, pink. *Filaments* and *styles* faintly pink. *Anthers* greenish. *Primordial fruit-stalk* short, shorter than the sepals, and as well as the terminal fruit-stalks of the branches shorter than the lateral stalks, which are erect-patent.

The prickles on the panicle of this plant are often exceedingly strong. One of the specimens of Leighton's *R. carpinifolius* has the under side of its leaves felted. The description of *R. carpinifolius* in the *Flora of Shropshire* was drawn up from a comparison of that in *Eng. Bot. Suppl.* (of a plant now combined with *R. Grabowskii*) with a specimen named *R. carpinifolius* by Borrer, but which is now believed to belong to *R. Lindleianus*. It is not therefore wonderful that Leighton's description does not agree with the specimens published as the *R. carpinifolius* of his *Flora* in his *Shropshire Rubi*. But I have strong reason to believe that that specimen really represents the plant which he always knew by the name of *R. carpinifolius.*

It has often been suspected that this may be the *R. horridus* (Hartm.) of Arrhenius. That plant is stated to have a very hairy stem, leaves nearly always ternate and the terminal leaflet subovate, sepals exceedingly prickly throughout, whilst those of our plant have only a few aciculi at the base. It seems to be confined to Scandinavia. My specimens of it have no good barren stem. See *R. diversifolius* for further remarks upon Hartmann's plant.

γ *pallidus;* aculeis aciculis setisque validis sed paucioribus, *foliolis subtus mollibus* subtomentosis in venis pilosis, *paniculæ sæpe angustæ* ramis brevibus corymboso-racemosis patentibus vel ramis axillaribus ascendentibus, *pedunculo terminali ramorum quam lateralibus sæpe longiori.*

R. *Koehleri* γ *pallidus* Bab. Man. ed. 5. 106; ed. 6. 115.

R. *pallidus* Weihe in Bluff et Fingerh. Compend. Fl. Germ. i. 682 (1825). Rubi Germ. 75. t. 29. Bab.! Man. ed. 3. 100; ed. 4. 103. Blox.! in Kirby, 42. Leight.! Fl. Shrop. 236. Lees! Malv. 52 (excl. var. β).

R. *Koehleri* Blox.! in Kirby, 41; Fascic. of Rubi (sp.). Lindl.! Syn. ed. 1. 94. Lees! in Steele, 56; Bot. Malv. 53.

R. *Koehleri* ε *fuscus* Bell Salt.! in Phytol. ii. 132; Bot. Gaz. ii. 127.

R. *fusco-ater* Lindl.! Syn. ed. 2. 93.

R. *Radula* δ *foliosus* Bab.! Syn. 24; Man. ed. 2. 103.

R. *fuscus,* Baker, Suppl. Fl. York. 64.

Stem arching rather highly at the base, afterwards prostrate, slightly angular below, strongly angular above, with many unequal hairs, setæ and aciculi. *Prickles* subpatent, rather less strong than those of *var. a,* but similar, as are also the aciculi and setæ. *Leaves* quinate. *Leaflets* stalked, flat, doubly and patently dentate, pilose above, pale green with long soft hairs on the veins beneath, and often (especially towards the upper end of the stems) with a thin coat of felt between them; all obovate-cuspidate; terminal broad, sometimes roundish; petioles (which are not furrowed above) and midribs with strong hooked prickles beneath; stipules linear-lanceolate.

Flowering shoot from fuscous scales, very hairy, setose and aciculate. *Prickles* slender, large-based, declining. *Leaves* ternate. *Leaflets* rather finely toothed; basal nearly sessile, broadly lanceolate, unequal-sided and often lobed on the outer edge, or roundish and cuspidate; terminal broadly obovate or roundish, subacuminate. *Panicle* long, leafy, with short axillary corymbose or racemose few-flowered branches; ultra-axillary branches few and short. *Sepals* aciculate, setose, hairy, felted, lanceolate-acuminate, with a slender rather leaflike point. *Petals* ovate-spathulate, distantly serrate, pink. *Filaments* pink. *Anthers* fuscous even in the unopened bud. *Styles* greenish. *Primordial fruit-stalk* short; but the terminal peduncle of the upper branches is usually longer (often considerably) than the lateral ones, which latter are commonly erect-patent.

The several plants which I have included under the name of *R. Koehleri* have been considered as distinct species by high botanical authorities; but they seem to be so closely connected by intermediate forms as to constitute one species. It is often difficult to determine under which of the named forms some specimens should range. They are well marked by the numerous strong and very unequal prickles on the barren shoots, of which the smaller so merge in aciculi and stiff aciculiform setæ, those in true setæ, and these last in hairs, that it is impossible to say where one of those forms of armature begins and another ends. Although the prickles are always abundant their number varies considerably : the stem of the typical plant is sometimes completely covered with their enormous bases together with those of the aciculi and setæ. When that is the case the tubercles are very much compressed and extended along the stem. In the *R. pallidus* the tubercles are sometimes considerably separated, and then assume a rather oval form.

In the typical plant the underside of the leaflets is quite

18

devoid of felt, is rough to the touch, and the hairs upon its veins are few in number and short ; the panicle is rather open, pyramidal and truncate, most of its branches being longish, corymbose and patent; the terminal flower of the whole panicle and of each of the branches is very shortly stalked, whilst the lateral flowers have much longer stalks; the rachis, branches and peduncles are nearly always very thickly armed with long slender prickles and aciculi, and have also many hairs and setæ. The petioles are furrowed above. The filaments are white, and the anthers pale yellow.

The variety which I call *infestus*, the *R. carpinifolius* of Leighton, has the under side of its leaflets often slightly felted, soft to the touch, with many hairs on all the veins ; the panicle is very prickly, broad, with an almost hemispherical top, and short mostly axillary branches; the prickles of the panicle, as also those of the stem, are usually short, thick, very much compressed, and falcate, or very greatly declining. The terminal peduncle of the panicle, and of each branch, is shorter than the lateral ones, which are here and in *R. pallidus* erect-patent, not divaricate, as in true *R. Koehleri*. The filaments are faintly pink, and the anthers greenish.

In *R. pallidus* the underside of the leaflets is usually furnished with a very fine coat of felt, and the veins bear many long hairs; therefore it is soft to the touch. The panicle is usually close, from the shortness of its branches, and generally narrows towards the top. The terminal flower of the panicle and of each of the branches is usually furnished with a longer stalk than the lateral flowers; and the prickles of the rachis and peduncles are rarely so abundant, and are nearly always stronger than those of *R. Koehleri*. The branches of the panicle have a tendency to become corymbose with divaricate branchlets in *R. Koehleri*, whilst those of *R. pallidus* are usually small racemes. The petioles of the latter are not furrowed, the filaments are pink, and the anthers fuscous.

Mr Borrer referred the *R. affinis* of Smith (*Eng. Fl.* ii. 405) to *R. Koehleri* γ *pallidus;* but the specimen so named and identified with Smith's plant in his Herbarium is *R. rosaceus.* It was gathered at Woodmancote near Henfield. It appears probable therefore that Borrer's *var. pallidus* includes my *R. rosaceus.* Mr Edw. Forster also considered the typical form of *R. Koehleri* to be the *R. affinis* of Smith, but it seems nearly impossible that Smith can have had one of the *Glandulosi* before him when drawing up his description of *R. affinis.*

The *R. fusco-ater* of Lindley's *Synopsis*, ed. 2, is shown to be *R. Koehleri* γ *pallidus* by the specimen from the Hort. Society's Garden in Herb. Borrer.

My *R. fusco-ater* γ *aculeatus* seems to be properly referred to *R. pallidus.* Its stem and flowering-shoot have very few hairs or hair-like setæ, but an abundance of aciculiform ones and aciculi and prickles. All nevertheless standing quite separate from each other, and having much less compressed bases. Its leaves are whiter beneath and more felted. Its panicle is more open and more pyramidal, and the terminal flowers are on shorter stalks.

A plant gathered by Mr H. C. Watson, at Chessington in Surrey, has precisely the same kind of prickles as *R. Koehleri*, and perhaps about as many of them, but they are very small and short, and therefore leave much of the cuticle naked. The only leaf which I have seen has four leaflets; the two on one side being those of a palmate leaf, both stalked, and the basal one directed backwards so as to be quite clear of the other; on the opposite side the single leaflet is dilated externally but not lobed; they are very slightly felted beneath, the veins are scarcely at all hairy,

and the prickles on the midrib and petiole are few and
weak. The panicle has a remarkable appearance; for its
branches (which are few) are erect, the uppermost alone
spreading so as to be erect-patent, and their lengths are such
as to place the flowers in an irregular convex corymb. On
my specimen there are only two branches which do not form
part of this corymb and which are not themselves corymbose;
their lower half is long and naked and the upper forms a
raceme of flowers. It seems not impossible that this may be
a state of *R. pallidus*, but my materials are not sufficient
from which to form a satisfactory opinion.

The *R. glandulosus* of Smith is very different from that
of Bellardi, and is unquestionably referable to *R. Koehleri*
a verus. It is the var. *cuspidatus* (an ill-chosen name) of my
Synopsis. Its leaflet is of an unequally rhomboidal form
(the lower half of the rhomb being longer than the upper),
with its upper part very regularly narrowed to the point,
but having its edge lobate-serrate; the lower part likewise
narrows gradually until close to the base, where it is rounded
and slightly notched. The upper part of the leaflet in true
R. Koehleri is often very similar, but the tip projects slightly
more from the general outline; also, the base is rather
broadly or truncately cordate. The panicle of Smith's plant
is almost exactly that of *R. pallidus*. This plant seems to
lie between those two marked forms of the species, and has
helped to convince me of their specific identity. It is certain that this is the *R. glandulosus* of Smith, for Mr D.
Turner (who originally sent it to Smith from Rydal in
Westmoreland) identified with his plant the specimens
gathered at the same place by Mr Borrer, who kindly presented some of them to me. His words were that "*R.
glandulosus* (Sm.) is wholly this plant of this place." The
other plants which I placed under var. *cuspidatus* approach
more nearly to the true *R. Koehleri*.

R. echinatus of Lindley and of Leighton, if determined by the specimens named by Lindley, is a form of *R. Koehleri a verus*, having an obovate-oblong acuminate leaflet, which is doubly or often rather patently serrate through more than its upper half. The panicle when pressed has a singular appearance owing to the long simple divaricate stalks of the lateral corymbs, and is very similar in look to that of some forms of *R. glandulosus* (Bell.). But if it is determined by the authentic plant in the Horticultural Society's garden, from which there is a specimen preserved in the Herb. Borrer, it is the *R. rudis* of Weihe, and of Lindley's *Synopsis*, ed. 2.

The specimen named *R. pallidus* by Nees for Leighton has ternate leaves with very coarsely, but often slightly doubly, serrate leaflets; the lower are strongly lobed on the outer edge and all are glabrous beneath, with the exception of a few distantly scattered hairs on the veins. The prickles and other arms of its stem are few in number. In all probability it was taken from a plant which grew in a shady place, and Nees von Esenbeck has correctly named it, notwithstanding its considerable difference from the plate in *Rubi Germanici*.

There does not seem to be much cause for doubting the identity of *R. Koehleri* and *R. pallidus* with the plants so named in the *Rubi Germanici*, although in neither case does the plate exactly represent our plant.

What I continue to call *infestus* does not quite agree with the *R. infestus* (Weihe), which has roundish-cordate terminal leaflets, much smaller prickles on the panicle, and much longer stalks to its terminal flowers. My plant is certainly the *R. carpinifolius* of Leighton; although the specimen so named for him by Borrer, which is now before me, is not the same; nor is it the plant of *English Botany* (which I now refer to *R. Grabowskii*) from which Mr Borrer himself pointed out its differences. I now think the plant named by Borrer is *R. Lindleianus.*

A beautiful plant, received from Mr Baker as *R. fuscus?,* and gathered at Laskill bridge in Ribsdale, much resembles some states of *R. pallidus.* Its chief differences are found in the smaller quantity of hair upon its barren stem, the rather hard feel of the under side of the leaves owing to the total want of felt and fewness of the hairs there, and the much more cylindrical and less hairy panicle. It seems to be one of the links connecting together the plants now placed under *R. Koehleri.*

Habitat.—Hedges and thickets. July, August.

Area.—1 2 3 4 5 . 7 8 9 10 11 12 13 . 15 16 . . 19 . . . . 30.

Localities of a.—i. Linton, *N. Dev.;* Tresco, Scilly, *W. Corw.* (Townsend!)—ii. St Leonard's Forest, *W. Suss.* (Borr.!). —v. *Worcester;* Shrewsbury, and near Wrexham, *Salop;* Cheltenham, *E. Glouc.* (Notcutt!).—vii. Bangor, *Caern.*—ix. Knutsford, *Chesh.*—x. Thirsk, *N. E. York.*—xii. Stock Gill, Ambleside and Rydal, *Westm.;* Douglas, *Isle of Man.*

xix. Killarney, *S. Kerry.*—xxx. Carumoney, *Antrim* (Tate!).

Of β.—iii. Claygate, and St Ann's hill, *Surr.*—v. Wyck, *W. Glouc.;* Bromsgrove Lickey, *Worc.;* Sharpestones hill, *Salop.*—vii. Llanberis, *Caern.*—x. Hebden bridge, *S. E. York.;* Cleadon and Thirsk, *N. E. York.*

xiii. Gourock, *Renf.*

Of γ.—i. [Leigh woods, Bristol, *N. Som.*—ii. Balcombe, *E. Suss.* (Mitten!); Henfield, *W. Suss.* (Borr.!).—iii. Chessington, *Surr.;* Easney Park and Oxhey wood, *Herts.* (Hind); Trent Park, *Middl.* (Hind!).—iv. Fakenham, *W. Norf.;* Balsham, *Cambr.*—v. Coleford and Lydney, *W. Glouc.;* Chepstow, *Monm.;* Broad Heath and Cowleigh Park, *Worc.;* Berwick and Almond Park, and Wrekin,

Salop; Needwood, *Staff.;* Ross, *Heref.* (Purchas!)—vi. Tenby,
Pemb.—vii. Llanberis, *Caern.;* Capel Garmon, *Denb.*—viii.
Twycross, *Leic.*—ix. Bradbury wood (Blox.!), Beeston Castle
(Bell Salt.!), Bowden (G. E. Hunt!), *Chesh.*—x. Sheffield,
S. E. York.; Thirsk and Bilsdale, *N. E. York.*—xi. *Durham.*
—xii. Keswick, *Cumb.* (Hort.!)

xiii. Jardine Hall, *Dumfr.*—xv. Inverarnan, *W. Perth.*—
xvi. Arran, *Clyde Isles.*

xix. Killarney, *S. Kerry.*

31. R. fusco-ater Weihe.

R. caule arcuato-prostrato angulato hirto, aculeis inæqualibus e basi magna compressa paululum declinatis, aciculis validis inæqualibus setisque multis, *foliolis* irregulariter vel subduplicato-dentatis *supra planis* subtus viridibus pilosis, foliolo terminali late cordato-obovato acuminato vel subcuspidato *infimis petiolatis intermediis incumbentibus*, paniculæ longæ subpyramidalis infernè folosæ ramis patentibus corymbosis vel ramis axillaribus erecto-patentibus racemosis, aculeis multis inæqualibus iis in medio caulis florentis quam reliquis majoribus, pilis setis aciculisque multis inæqualibus, *sepalis* ovato-attenuatis setosis aciculatis *patentibus vel ad fructum adpressis.*

R. fusco-ater Weihe in Bluff et Fingerhut Compend. Fl. Germ. i. 681 (1825). Rubi Germ. 72. t. 26. Blox.! in Kirby, 40. Bab.! Man. ed. 5. 106; ed. 6. 115.

R. scaber Lees! Malv. 53.

R. hirtus Lees! in Steele 55. Bell Salt.! in Bot. Gaz. ii. 127 (excl. β).

Stem arcuate-prostrate, usually very prickly, very hairy, angular; sometimes the old stems become nearly naked from the hairs, setæ and aciculi being deciduous. *Prickles* very unequal, declining, from long compressed bases. *Aciculi* and *setæ* many, unequal; some of the setæ only differ from the more slender aciculi by having glandular heads. *Leaves* quinate-pedate, slightly convex. *Leaflets* all overlapping, dull, slightly convex, smooth and pilose above, green and hairy on the veins beneath, doubly dentate, wavy at the edge; basal leaflets oblong, nearly but not quite sessile; intermediate broadly elliptic, shortly stalked; terminal broadly cordate-ovate or broadly cordate, shortly stalked; petioles (which are flat above) and midribs with strong hooked prickles beneath; stipules linear.

Flowering shoot from fuscous scales, very hairy, setose, aciculate, with many slender declining or deflexed prickles. *Leaves* ternate. *Leaflets* nearly equal, coarsely and doubly dentate; basal ovate with the outer side gibbous and often lobed; terminal broadly oval, acuminate, narrowed to the base. *Panicle* long, rather pyramidal; axillary branches about equalling the leaves, racemose; ultra-axillary few-flowered, divaricate, subcorymbose; rachis and peduncles with many purple unequal aciculi and setæ. *Sepals* ovate, acuminate, felted, hairy, setose, aciculate, clasping the fruit. *Petals* obovate, sometimes much narrowed below, pink. *Filaments* pink. *Anthers* yellow. *Styles* red. *Primordial fruitstalk* as long as the sepals; those of the branches longer than the lateral ones. *Nut* half-ovate.

It is nearly certain that our *R. fusco-ater* is identical with the plant called by that name by Weihe. It is quite distinct from the *R. fusco-ater* of most British botanists. But we must mention that the authors of the *Rubi Germanici* state that the prickles of the *R. fusco-ater* are not much dilated at the base, whereas on our plant the dilatation is often very remarkable; that the petals of their plant are broad, and such petals may be found on English examples of this species; that the anthers are "intense purpurea;" and the sepals "post anthesin reflexis." Should these differences prove to be real, and be thought of sufficient weight to separate the plants, our bramble will require a new name. There is room for doubt concerning the prickles, for I suspect that the artist has not always correctly represented their mode of springing from the stem; and the colour of the anthers may have resulted from age, for I have found that it very frequently changes to a dark tint after the pollen has fallen. The direction of the fruit-sepals seems uncertain, if not variable, in their plant. As our plant accords very accurately with their plate and description in other respects, there is reason to believe in the identity of the species.

Plants found by Mr Newbould near Sheffield differ slightly from the other specimens. Their stems are less prickly, having the large prickles relatively more conspicuous. Mr Bloxam's specimens differ amongst themselves; those published in his *Fasciculus* have very prickly and nearly glabrous stems, bearing very few aciculi, and still fewer setæ; others kindly sent to me are as setose and aciculate as those gathered by Mr Mathews of what I consider to be the typical plant. In both of his plants the panicle is shorter, less pyramidal, and more open at the top than in those of Mr Mathews.

The plant which grows at Henfield is more like the plate in *Rubi Germanici* than any other specimens which I have seen; but such is the deciduous tendency of the arms, that the old stems very closely resemble those of the *Caesii.* Indeed the denuded state is so like *R. Balfourianus* that I am unable to point out any satisfactory character by which to distinguish the plants; although I believe them to be quite distinct. The young and adult stems of *R. fusco-ater* clearly show that it belongs to the *Koehleriani* and present no trace of the glaucous bloom which is usually present on those of the *Caesii.* The aciculi and setæ on the stems of *R. Balfourianus* can rarely be called abundant. It is only by a combination of characters, not one of which perhaps is constant, that we can distinguish plants the typical forms of which are as different as any two species of fruticose bramble. This is one of the cases by which we are taught to make allowance for those botanists who fancy, most erroneously as I believe, that our species all run into each other to such a degree as not to allow of their separation. A plant which Mr Briggs informs me is abundant near Plymouth very closely resembles that found at Henfield by the late Mr Borrer; but its leaves are (I suppose) all ternate, whereas those of the Henfield plant are nearly always quinate. The adult stems also are less but similarly armed. The specimens from Henfield are in flower; those from Plymouth in fruit, so that they cannot be quite satisfactorily compared. Mr Bloxam separates the Plymouth plant from other brambles, and names it *R. Briggsii* (Seem. Journ. of Bot. vii. 3. t. 88). Others found at the Slate Houses, Henfield, were named *R. Bakeri* by Mr Bloxam, but they differ greatly from the state of *R. villicaulis* similarly named by him. They have the terminal leaflet cordate, the basal leaflets sessile, the panicle close and short, and the sepals adpressed.

The *R. scaber* of Lees (from Storrage hill), seems to be a state of this species, but differs slightly from the type. Its stems have few hairs, but plenty of very short aciculi and setæ in addition to the large thick-based ones. The under side of its leaves is covered with very minute hairs. The hairs on the stem seem to be often deciduous, and may have been so in this case.

R. diversifolius will be seen by the description to differ considerably from this plant. Its leaves are rugose above, the basal leaflets are sessile, the panicle leafy almost to the top, and the sepals are unarmed. Indeed it is far more different in appearance than can be shown by description.

Some specimens (which I refer to *R. fusco-ater*) show a slight tendency to have the under side of their leaves felted, but it can only be detected by the use of a powerful magnifying glass. When such is the case with the leaves of the stem those of the flowering shoot are usually (perhaps always) densely felted beneath. The lower leaflets of these plants require attention, for it is out of my power to ascertain from the dried specimens what was their direction when the plant was alive, and therefore cannot be certain that they were incumbent. Also, their calyx seems less inclined to clasp the fruit, even if its tendency was not to be reflexed. It is therefore quite possible that they are misplaced here.

I have one specimen of a magnificent plant found by Leighton in a hedge by the road-side between Shrewsbury and Berwick which agrees better with *R. fusco-ater* (as it was named by Mr Bloxam) than with any other bramble that is known to me. It has a dense nearly cylindrical panicle which is leafy nearly to its top, and has short many-flowered truly corymbose branches. Its basal leaflets (on the stem) are very nearly sessile.

Habitat.—Hedges and heaths. July, August.

Area.— . 2 . . 5 . . 8 . 10 11.

Localities.—ii. Henfield, *W. Suss.*—v. Near Tintern, *W. Glouc.;* near the fir trees on the top of Bromsgrove Lickey, on old Storrage hill, and in Wyre Forest, *Worc.;* Sutton Park near Birmingham, and Wyken lane, *Warw.;* Bog at Almond Park, *Salop.*—viii. Ashby de la Zouch and Twycross, *Leic.*—x. Loxley near Sheffield, *S. W. York.;* Goldsborough, *Mid. W. York.;* Wass, *N. E. York.*—xi. Whitley, *Northumb.*

32. R. diversifolius Lindl.

R. *caule* arcuato-prostrato angulato *sparsim piloso,* aculeis inæqualibus è basi compressa subpatentibus, aciculis setisque multis inæqualibus, *foliolis* sæpè regulariter (apicem versus subduplicato-) dentatis ad marginem undulatis *supra rugosis* subtus pallide viridibus pilosis et *sæpissime* tomentosis, foliolo terminali late cordato-obovato-acuminato *infimis sessilibus intermediis incumbentibus, paniculæ longæ fere ad apicem foliosæ* ramis erecto-patentibus subracemosis, aculeis in medio caulis florentis quam reliquis majoribus, pilis setis aciculisque brevibus æqualibus, *sepalis ovatis acutis* tomentosis setosis *patentibus fructuive laxe adpressis.*

R. *diversifolius* Lindl.! Syn. ed. 2. 94 (1835). Bab.! Man. ed. 5. 106; ed. 6. 115.

R. *fusco-ater* a Bab.! Syn. 25; Man. ed. 2. 103; ed. 3. 101; ed. 4. 103. Bell Salt.! in Phytol. ii. 132. Leight.! Shropsh. Rubi, 18 (sp.).

R. *dumetorum* Leight.! Fl. Shrop. 237.

R. *dumetorum* δ *ferox* Lees! in Steele, 54.

R. *ferox* Weihe in Boeningh. Prod. Fl. Monast. 153.

R. *nemorosus* δ *horridus* Bab.! Syn. 33; Man. ed. 2. 107.

R. *Radula* Leight.! Fl. Shrop. 232 (in part). Lindl.! Syn. ed. 2. 94 (in part).

R. *Koehleri* δ *fusco-ater* Bell Salt. in Fl. Vect. 159.

R. *Schleicheri* Leight.! Fl. Shrop. 237. Godr. Fl. Lorr. ed. 2. 234. Bell Salt. in Phytol. ii. 131. Bab.! Syn. 31; Man. ed. 2. 106.

R. *entomodontos* Müll.! in Billot Annot. 292 (?) (1861).

19

Stem arcuate-prostrate, angular, sometimes furrowed, very hairy when young. *Prickles* unequal, slender, patent or very slightly declining, with very long compressed bases, the larger ones chiefly seated on the angles of the stem. *Aciculi* and *setæ* many, rather strong, unequal, seated on tubercles. *Leaves* quinate-pedate. *Leaflets* broad, dark green, opaque, rugose and pilose above, pale green, pilose and felted beneath, nearly regularly dentate, or slightly lobate-dentate towards the tip; basal sessile, oblong, overlapping the intermediate pair; intermediate shortly stalked, oblong-obovate, unequal-based; terminal roundly cordate-obovate, acuminate or cuspidate, shortly stalked; or rarely the basal and intermediate are combined into one leaflet which is largely and deeply bilobed; petioles and midribs with rather few short rather slender unequal hooked prickles beneath; stipules narrowly lanceolate.

Flowering shoot from brownish silky scales, straight except near the top, where it is slightly wavy. *Prickles* slender, from long compressed bases, increasing in length from the base of the shoot to about its middle, then decreasing gradually to the summit. *Aciculi* and *setæ* rather plentiful, short. *Hairs* fascicled, interlacing, and often rather adpressed. *Leaves* ternate. *Leaflets* nearly equal, pilose above, hairy and felted beneath; basal very unequal-sided, the outer side being half rhomboidal; terminal obovate, acute, somewhat wedgeshaped below; uppermost floral leaves simple, either 3-lobed or broadly oval. *Panicle* long, leafy nearly to the top, with very short axillary few-flowered subracemose branches often springing from every axil of the shoot; sometimes a few of the branches are rather longer, although still short, and become secondary and leafy panicles. *Sepals* ovate, acute or with slender points, whitish, setose, aciculate, felted, erect-patent and slightly clasping the fruit or loosely reflexed from it. *Petals* ovate, slightly notched, clawed, white. *Stamens* and *styles* yellowish. *Primordial fruit-stalk* shorter than the sepals, bearing a fruit, which is often small, from few of the large black drupes ripening.

This plant seems to approach the *R. Wahlbergii* (Arrh.), but the authentic specimen (Fries, *II. N.* ix. 49) is different, and will be noticed under *R. corylifolius* γ *purpureus*. Arrhenius appears by his description to intend to convey the idea (although he does not actually say so) of a plant wanting setæ on its stem, but having them on its flowering shoot. As it is quite impossible for that botanist to have overlooked the abundant glands which tip the smaller aciculi of *R. diversifolius* as well as the plentiful setæ, it appears certain that it is not *R. Wahlbergii*. It may probably be the *R. nemorosus* c. *ferox* (Arrh.), and the variety of *R. dumetorum*, so named on table 45 B of the *Rubi Germanici;* but the plant there represented is far more prickly throughout than *R. diversifolius*.

That our present plant is the *R. diversifolius* (Lindl.) is shown by his own authentic specimens and by the remark in the second edition of his *Synopsis* (94). That he also gave the name of *R. Radula* to it is similarly shown by his specimens now before me. Mr Baker gathers a form of *R. diversifolius* abundantly in N. E. Yorkshire which has no felt beneath its leaves, but seems to agree in all other respects with the true plant. He states that sometimes there is a little white felt on the leaves. *R. Schleicheri* of Leighton appears never to have any felt, but typical *R. diversifolius* almost always has a considerable quantity; nevertheless I believe the two are states of one species. Mr Baker's specimens, although without felt, are more near to the type than to *R. Schleicheri*. M. Genevier identifies this plant with the *R. horrefactus* Müll.

The specimens marked "*R. dumetorum* W. and N. var. β

nemorosus ad *var. a ferocem* accedens, si calyces fructus sint erecti" by Nees, which are mentioned in Leighton's *Flora* (238), were referred by Borrer to *R. dumetorum* and recognised by him as the *R. diversifolius* of Lindley's latter opinion. They certainly are the true *R. diversifolius* as ultimately understood by Lindley, and now recognised as such by me. Also other specimens sent by Leighton to Borrer and Lindley, and returned named by them, seem to belong to *R. diversifolius.* They are marked Nos. 25 and 26, and considered as "undoubtedly *R. nemorosus*" by Leighton, "*R. Radula*" by Lindley, and "*R. cæsius*" by Borrer. I believe them to be states of *R. diversifolius* with fewer prickles than usual, and smaller and less compound panicles; they are probably the shoots of young plants. Another specimen, No. 16, was considered by Nees von Esenbech to be "*R. dumetorum β nemorosus.*" Leighton believed it to be the same as Nos. 25 and 26, but I think that it belongs to the *Radulæ.* Its stems are young and not in a satisfactory state for examination, but its leaves and panicle seem to prove that it is a state of *R. Hystrix.*

The *R. Schleicheri* (Leight.) appears to be a state of this species. It agrees far more nearly with *R. diversifolius* than with any other of our plants. The chief differences seem to be that the panicle is usually furnished with longer and more spreading branches, the leaves are nearly or quite devoid of felt, being only densely hairy on the veins beneath, and the terminal leaflet is rather longer in proportion to its breadth. I have no doubt of its being rightly placed amongst the *Koehleriani;* although it often closely resembles *R. tuberculatus,* which belongs to the *Cæsii. R. tuberculatus* has a decided bloom, and very inconspicuous, short and nearly equal aciculi and setæ upon the stem, prickles springing from large oval depressed tubercles, a terminal leaflet, which is usually much broader and more cordate at the

base, and the upper part of the flowering shoot furnished
with the longest prickles (their size and length decreasing
gradually from a short distance below the top to the base).
Apparently the *R. Schleicheri* (Weihe), if we are to judge
from the plate in the *Rubi Germanici*, is nearly allied to, and
may be identical with, our *R. diversifolius*. It manifestly
belongs to the *Koehleriani*, and is placed close to *R. Koehleri* by Dr Metsch. Such also seems to be the case with
the *R. Schleicheri* of Godron, who states that his plant is
the *R. glandulosus* of Schleicher's *Plantæ exsiccatæ;* but it
is doubtful if as much can be said of that described by Boreau, who seems to have drawn up his account of it from a
combination of the description and figure in the *Rubi Germanici.* The specimens sent by Leighton to Nees v. Esenbeck (of which I have two examples before me), named
R. Schleicheri by him, and therefore so called in Leighton's
Flora, belong to *R. diversifolius.*

On the other hand Weihe's description of *R. Schleicheri*
in Bluff and Fingerhuth's *Compendium,* that in *Rubi Germanici,* that by Metsch in the *Linnæa,* and by Reichenbach
in the *Flora excursoria,* all seem to refer to quite a distinct
plant from our *R. diversifolius.* The author of the description in *Rubi Germanici* remarks: "aculei majores adunci,
minores reclinati, omnes autem conferti lataque basi caulem
quasi tuberculis exasperatis," by which I understand an armature such as is found amongst the *Radulæ,* and very
different from that represented on the plate in the same
work; and yet the author of the description adds in a note
that the plate is "satis fida." The large prickles represented on that plate are connected closely by those of intermediate sizes with the smaller prickles, those similarly with
the aciculi, and the latter with the setæ and hairs, as is the
case in all plants belonging to the *Koehleriani;* the prickles
and aciculi are also, as in that group, subulate in shape.

Our *R. diversifolius*, in its more prickly state, agrees very fairly with the plate in the *Rubi Germ.*, and in its least armed state with the specimen published by Billot (*Fl. Gal. et Germ.* No. 2451), and in a rather intermediate condition with those named *R. Schleicheri* for Leighton by Nees. Leighton describes the prickles as "scattered, very unequal, diminishing insensibly into setæ, straight and horizontal or slightly recurved," by which latter word he seems to mean "declining," which is compatible with absolute straightness.

From all this it will be seen how difficult it is to determine to what plant the name *R. Schleicheri* belongs. Weihe is its original author, and we may conclude with almost certainty that his plant is not our *R. diversifolius* nor the *R. Schleicheri* of Nees v. Esenbech. It is only by supposing that certain plates in the *Rubi Germanici* were prepared under the superintendence of one of the authors of that great work and the descriptions written by the other, that we can account for the differences which exist between them. In the present instance Weihe (who certainly ought to be followed in this case) seems to have named specimens in accordance with the description, and Nees from their agreement with the plate. And the difficulty is increased by all authors having quoted the plate as representing their plant, which we now see to be an impossibility. Dr Metsch remarks that *R. Schleicheri* as understood by him, and as intended (I believe) by Weihe, is distinguished from all others known to him by its "ternate green leaves, numerous strong much hooked prickles which have conspicuously thickened bases, which give a peculiar tubercular (höcheriges) appearance to the stem." It is difficult to tell exactly what is intended by this phrase, but I think that he had a structure like that of the *Radulæ* in view.

R. horridus (Hartm., Arrh.), of which I possess two Ostrogothic specimens, very much resembles *R. diversifolius;*

but Arrhenius says, and the specimens confirm him, that it has decidedly falcate prickles and ternate leaves on both shoots, leaflets that are ovate or roundly-ovate: in all these respects differing from *R. diversifolius.* He also adds that the barren stem is not glandular; but the specimens bear plenty of setæ. The panicle of these Swedish specimens is densely covered with long patent hair, and the sepals are exceedingly prickly.

R. entomodontos (Müll.! in *Bill. Annot.* 292), *R. Schleicheri* (Bill.! *Fl. Gall. et Germ. exsic.* No. 2451 sp.) approaches *R. diversifolius* very closely, but has a nearly leafless broad rather dense cylindrical panicle. The *R. viretorum* (Müll.! *Versuch,* 202. *Wirtg. Herb. Rub.* No. 186. sp.) also nearly approaches the *R. diversifolius,* and might perhaps be safely referred to that species. It scarcely differs except by having a broader top to its panicle and no felt beneath its leaves.

The application of the name, *R. diversifolius,* to *R. leucostachys,* in the first edition of Lindley's *Synopsis,* was most unfortunate. As has been already stated (p. 117) it was the cause of much difficulty, and would have justified the total neglect of the name. But as I believe the present plant to be that really intended by Lindley, it seems better to retain it than to add another to the long list of synonyms. Mr Borrer remarked upon the specimen, named by Lindley and submitted to him by Leighton, "if this is *R. diversifolius* (Lindl.) the Professor may well criticise my inclination to unite that species with *R. leucostachys;* but I have a very different thing from the garden of the Horticultural Society as from the authentic bush of *R. diversifolius.*"

I have specimens of a plant gathered at Henfield, by Mr Borrer, which are much like *R. diversifolius,* but nevertheless differ considerably from the species. The leaflets are very broad and the terminal one almost round with a small cusp and cordate base; their under side is not felted, but is sometimes so thickly covered with hairs (all on the veins) as to seem so at the first view, whilst other leaves are nearly naked beneath. I have seen nothing quite like this, and as it has only been found in one place it must be left for future consideration. It seems to be the *R. horrefactus* (Müll. *Mon.* 179), as it agrees well with specimens from Sheen Common in Surrey, and from Cleves near Thirsk in N. E. Yorkshire, which were so-named by M. Genevier.

The Rev. A. Bloxam sends specimens of a plant, gathered at Hutton near Waith in Yorkshire, which he thinks may be *R. apiculatus* (Weihe). As far as I can judge from them it is nearly allied to *R. Koehleri* and *R. diversifolius.* Either the hairs and setæ are very few in number or very deciduous. Aciculi are tolerably abundant, and when broken leave the peculiar short pyramidal base which is characteristic of the *Koehleriani.* The only leaves that I have seen are ternate : the lateral leaflets being unequal-sided or somewhat lobed (clearly consisting of two leaflets cohering): the terminal leaflet is oval, but slightly broadest just above the middle, cordate-based, acuminate. The under side of the leaflets is hairy and grey-felted ; the edge simply dentate below, rather doubly towards the tip, and then the main teeth are patent or even reflexed ; all the teeth are strongly apiculate. In all these respects it agrees sufficiently with *R. Schleicheri* of Leighton, which is a form of *R. diversifolius.* Those plants have a narrow open panicle which differs greatly at first sight from that of *R. diversifolius ;* nevertheless there seems really no material difference between them. Here the floral leaves are smaller, although mostly exceeding the axillary racemes. These racemose or even paniculate branches bear more numerous flowers and are rather more patent than those of *R. diversifolius.* On the whole I think that Mr Bloxam's plant is a form of *R. diversifolius*, although it may also be *R. apiculatus* (Weihe).

Mr Syme (*E. B.* iii. 184) would seem to refer the whole of my *R. Radula* γ *denticulatus* doubtfully to the *R. apiculatus.* In this I consider him to be wrong. The true *var. denticulatus* seems to me to be certainly a form of *R. Radula,* although not a very common one. He is I believe correct in saying that Mr Bloxam's *R. apiculatus* (MS.) can hardly be joined to *R. Radula,* and correctly quotes my opinion that it (not the true *v. denticulatus*) is nearly allied to *R. diversifolius.* It does not appear clearly from Mr Syme's words that he has ever seen the plant from Waith, and I have not seen that gathered by Mr Bloxam near Sheffield. Mr Newbould's plant from near Sheffield is the true *v. denticulatus* of *R. Radula;* and if the specimen given to Mr Syme by that botanist has the armature of the *Koehleriani,* it is not the same plant as I received from him in the year 1846.

The specimen named *R. fusco-ater* by Dr Bell Salter for Leighton differs from that for which I am indebted to Dr Salter himself. The former is *R. diversifolius;* the latter may be a more than usually prickly state of *R. Balfourianus.*

The special characteristics of this species seem to be the very prickly stem with longitudinally flattened prickles, imbricate lower leaflets, and a panicle having a slightly wavy but very strong rachis and an abundance of short nearly equal axillary branches which always fall short of the leaves.

Habitat.—Hedges. Exceedingly abundant in some places, especially in the valley of the Severn in Montgomeryshire and Shropshire. July and August.

Area.—1 2 3 4 5 6 7 . 9 10.

Localities.—i. St Mary's, Scilly, *W. Corn.* (Townsend!); Near Plymouth, *S. Dev.* (Briggs).—ii. Cockleton Bog, *Isle of Wight* (Salter!); Poole, *Dors.* (Salter!); Selborne, *Hants* (Salter); Henfield, *W. Suss.*—iii. Watford, *Herts;* Clapham,

(E. Forster!) and Sheen Common, *Surr.*—iv. Kingston, Caldecot and Hildersham, *Cambr.*—v. Shrewsbury, The Wrekin and Pattingham, *Salop;* Rugeley, *Staff.*—vi. New Radnor, *Radn.*—vii. Welshpool, *Montgom.;* Pen maen mawr, *Caern.* —ix. Bowdon, *Ches.* (G. E. Hunt!); Warrington, *S. Lanc.*— x. Thirsk, *N. E. York.*

33. R. Lejeunii Weihe.

R. caule arcuato-prostrato subangulato sparsim piloso et setoso, aculeis plerisque parvis nonnullis longioribus e basi longa compressa declinatis, *aciculis brevissimis*, foliis quinato-pedatis vel raro ternatis, *foliolis*
supra opacis pilosis subtus pallidioribus in venis tantum
pilosis apicem versus lobato-serratis *infimis petiolatis*
intermediis dissitis, foliolo terminali obovato-acuminato,
paniculæ latæ hirtæ foliosæ apice corymboso *ramis*
axillaribus *ascendentibus* subracemosis, aculeis tenuibus
declinatis, setis inæqualibus multis *sepalis ovatis* tomentosis setosis fructui laxe adpressis vel patentibus.

R. Lejeunii Weihe in Bluff et Fingerh. Compend. Fl.
Germ. 683 (1825). Rubi Germ. 79. t. 31. Bab.! Man. ed.
1. 97; ed. 5. 106; ed. 6. 116. Bell Salt.! in Phytol. ii. 135.
R. glandulosus β Lejeunii Bab.! Syn. 30; Man. ed. 2.
105; ed. 3. 102 ; ed. 4. 105.
R. Bellardi β Lejeunii Lees in Steele, 55.

Stem (arching slightly, afterwards nearly prostrate,?)
slightly angular. *Prickles* many, unequal, small, declining
from a long compressed base. *Aciculi* very short but strong,
springing from tubercles. *Setæ* and *hairs* few, short. *Leaves*
quinate-pedate. *Leaflets* serrate below, lobate-serrate in
their upper half, dull and hairy above, rather paler and
hairy on the veins beneath ; basal very shortly stalked,
lanceolate ; intermediate lanceolate-acuminate, rather unequal at their base; terminal broadly lanceolate-acuminate,
subcordate at the base; sometimes the basal and intermediate

of the same side combine to form one strongly-lobed leaflet; petioles and midribs with small hooked prickles beneath; petioles apparently not furrowed above; stipules slender.

Flowering shoot angular, armed like the stem. *Leaves* ternate. *Leaflets* obovate-oblong, subcuspidate, lobate-serrate towards their end, green on both sides, pilose above, rather paler and hairy on the veins beneath. *Panicle* open; axillary branches rather long but rarely exceeding the leaves, racemose-corymbose; ultra-axillary part short, with short patent corymbose branches; peduncles and branches with many unequal straight declining prickles, very many unequal setæ, of which the longest scarcely exceed in length the abundant hairs, a few aciculi, and a thin coat of felt. *Sepals* ovate, with a short linear point, green with a narrow white border, hairy, felted, setose, aciculate, patent or loosely adpressed to the fruit. *Primordial fruitstalk* about as long as the lateral ones, shorter than the sepals.

Dr Bell Salter's plant from Selborne agrees well with the plate of *R. Lejeunii* given in the *Rubi Germanici*. Mr Hind's plant has a narrow and more leafy panicle, but agrees with this species in other respects. My plant from Guernsey has broader leaflets, which are rather cuspidate than acuminate. In Mr Gibson's plant from Essex more of the upper part of the panicle is leafless, and there are more large prickles on the stem, but fewer small ones.

Dr Salter remarks that "*R. rosaceus* may be known from *R. Lejeunii* by the far greater abundance of glands [setæ] in every part, by the leaves being ternate instead of quinate-pedate, by the absence of tomentum from the panicle and by the greater length of the calyx." In all these respects my plant from Guernsey is rather *R. Lejeunii*, as I originally (*Prim. Fl. Sarn.* 32) supposed, than *R. rosaceus*, as it was afterwards (*Phytol.* ii. 133) named by Dr Salter. The plant from Guildford, Isle of Wight, named *R. rosaceus* when collected in company with Dr Salter, but afterwards corrected by him into *R. Lejeunii*, is in my opinion certainly *R. rosaceus*. He continued to call it *R. Lejeunii* as lately as the time (1856) when the *Fl. Vectensis* appeared, for he there states that it is the only form of *R. glandulosus* (under which he places it as a variety) "yet observed in the island." He was probably led to hold this opinion concerning the true name of the plant by finding in *Herb. Borr.* a specimen gathered at Vervier by Mr Woods in company with M. Lejeune, and considered as certainly *R. rosaceus*, became so-named by the latter botanist. It is exactly like *R. Lejeunii*, and has, even more decidedly than our plants, the armature proper to the *Koehleriani*. I do not find that Lejeune even published any plant as *R. rosaceus*.

Wirtgen (*Fl. der preussischen Rheinprovinz*, 158) places *R. Lejeunii* as a variety of his *R. vestitus* which he places between *R. scaber* and *R. thyrsiflorus*, and combines our *R. leucostachys* with *R. discolor*. I cannot agree with either of these arrangements. Our *R. vestitus* (and I think that of continental botanists) is certainly a state of *R. leucostachys* which itself seems abundantly different from *R. discolor*.

Garke (*Fl. v. N. und Mitt. Deutschl. ed.* 7. 125) considers *R. Lejeunii* as absolutely identical with *R. glandulosus* (Bell.).

I see no reason to doubt the correctness of placing this plant amongst the *Koehleriani*. The short conical remains of aciculi on its stem are exactly like those of other plants belonging to that group, and differ from the tubercles of the *Radulae*. The *Bellardiani* present no trace of either of these structures. *R. Lejeunii* seems to be quite distinct from all our other species.

A plant is given in Billot's *Flora Gall. et Germ. exsic.* (No. 970) as *R. Lejeunii* which is not the same as ours, nor, I fully believe, as that figured in the *Rubi Germanici*.

In the *Botany of Worcester* Mr Lees states that he still

20

considers his *R. Lejeunii* to be a variety of *R. glandulosus.*
This was my opinion in the earlier editions of my *Manual;*
but I now believe that I was then in error.

    Habitat.—Banks and hedges. July, August.

    Area.— . 2 3 . 5 . 7 . 9 10 . 12.

    Localities.—ii. Between Temple and Walmer, Selborne,
S. Hants (Salter).—iii. Oxhey wood, Watford, *Herts;* Bar-
rack wood, Warley, *S. Essex* (Hind!); Debden wood, *N.
Essex.*—v. Bog at Almond Park, *Salop.*—vii. Near Dinas
Dinorwig, *Caern.*—ix. Bowdon, *Chesh.* (Hunt!).—x. Aln-
wick, *Northumb.* (Baker!).—xii. Douglas, *Isle of Man.*

    On the slope of Fort George next the sea, *Guernsey.*

    b. Bellardiani. Folia ternata vel raro quinato-
pedata; foliola infima pedicellata intermediis dissita.
Aculei in caulium aciculatorum setosorum valde hirtorum angulis sæpissime congesti.

    The *Rubi Bellardiani* have very hairy stems with many
slender, weak and unequal aciculi and setæ, which do not
spring from tubercles and are persistent. The prickles are
rarely very large, often project very slightly beyond the
dense coat of hairs, and are confined to the angles of the
stem. The stems altogether want the filelike surface found
in the *Radulæ,* and also the very prickly armature of the
Koehleriani.

34. R. pyramidalis Bab.

R. caule subarcuato-prostrato tereti-angulato, *aculeis* multis *brevibus validis* e basi magna compressa valde declinatis deflexisve, pilis paucis, aciculis setisque multis subæqualibus, foliis 3-natis vel raro quinato-pedatis, *foliolis* irregulariter dentato-serratis subæqualibus *convexis* supra opacis pilosis *subtus* pallidioribus *pilosis*, foliolo terminali obovato-cuspidato, *paniculæ pyramidalis* infernè foliosæ apice et ramis racemosis tomentosis *rachide recta rigida,* aculeis tenuibus declinatis, pilis et setis inæqualibus multis, sepalis lanceolato-attenuatis fructui laxe adpressis.

R. pyramidalis Bab.! in Bot. Gaz. i. 121 (1849); Man. ed. 3: 101 ; ed. 6. 116 ; in Billot Annot. 135. Symes' Eng. Bot. iii. 188.

R. Güntheri β. pyramidalis Bab.! in A. N. H. ser. 2. ii. 40 (1848); Trans. Edin. Bot. Soc. iii. 59.

R. longithyrsiger Lees! MS. (1849).

Stem prostrate almost from its base and conforming itself to the inæqualities of the ground, rooting, angular, not furrowed, greenish purple. *Prickles* many, short, much declining, from a long compressed base. *Hairs* few. *Setæ* and *aciculi* many, nearly equal. *Leaves* ternate or rarely quinate-pedate. *Leaflets* of the ternate leaves nearly equal, green on both sides, dull with scattered hairs above, paler with yellowish hairy veins beneath, not felted, irregularly dentate-serrate, convex from the edges being bent downwards, obovate-cuspidate; basal unequal-sided; terminal slightly cordate at the base; the quinate leaves have the basal leaflets equally obovate; intermediate unequal-based and cuspidate, terminal obovate cuspidate; petioles (which are not furrowed above) and midribs armed beneath, similarly but less strongly than the stem; stipules linear.

Flowering shoot from brown scales clothed with ashy silky down, very hairy. *Prickles* rather many, short, a few longer and declining from long bases. *Aciculi* and setæ few, short. *Leaves* ternate, large, like those of the stem; upper floral leaves simple, ovate or cordate and lobed. *Panicle* very long; branches long, axillary, ascending, racemose, felted, the upper few-flowered and rather corymbose; top ultra-axillary, pyramidal, with rather long few-flowered or 1-flowered patent or divaricate branches, with many aciculi and setæ; whole panicle pyramidal, very stiff, very hairy, with slender straight prickles; general and partial rachis and peduncles nearly or quite straight. *Sepals* lanceolate, attenuate, long-pointed, felted, with a few purple setæ and a few purple aciculi. *Petals* distant, narrowly ovate, attenuate below, greenish white, often more than five in number. *Filaments* white. *Anthers* greenish. *Styles* pink at the base, becoming more pink after the petals have fallen, pale green towards their top; there is a broad clear flat space between the stamens and pistils. *Primordial fruit-stalk* as long as the calyx. *Primordial fruit* oblong, rather longer than the other fruit, closely surrounded by the calyx which is either patent or forced back by it. *Nut* roundly ½-ovate; inner edge quite straight until near the top; style subterminal.

This is one of the most beautiful, if not the finest, of our brambles; its panicle is often enormous, being several feet in length, and of a very markedly pyramidal form. Its lower panicle-branches are often very long, and resemble secondary panicles. Its narrow, distant but numerous, and

greenish-white petals are remarkable. The fruits are amongst
the best flavoured of those known to me. By some botanists
it has been supposed that this is the *R. thyrsiflorus* of the
Rubi Germanici; but that plant has quinate leaves, many-
flowered upper panicle-branches, and broad petals.

The Abbe Questier has issued specimens in Billot's *Flora
Gallica et Germanica exsiccata* (No. 2058) with the name
of *R. pyramidalis*, supposing them to be the same as my
plant. Unfortunately I have been the cause of this error,
by admitting, that the name of *R. pyramidalis* was appli-
cable to a specimen sent to me as such by Mr Questier.
The examination of the specimen contained in Billot's
collection, has now quite satisfied me that that plant is the
R. Güntheri of English authors, and probably the *R. cinerascens* of Boreau. It is perhaps not the same as another
specimen sent to me as *R. pyramidalis* by Questier; but of
that I am uncertain, although I feel no doubt of the latter
not being my *R. pyramidalis*. This error was corrected in
Billot's *Annotations*, p. 135.

It is worthy of remark that *R. pyramidalis* has leaves
which are devoid of felt, although it rejoices in the full
light of the sun; whilst *R. Güntheri*, which is usually an
inhabitant of shady places, is furnished with an abundance
of felt. This is strongly opposed to the idea of some botanists,
that *R. Güntheri* is the wood form of the same species as
R. pyramidalis; for in such cases it is always found that
the felt disappears as the depth of shade increases.

I have not seen the true *R. thyrsiflorus* in Britain;
Dr Godron combines it with *R. Güntheri* and *R. hirtus,*
but they seem to me to be abundantly distinct from what
we believe is the true *R. hirtus* (see *R. glandulosus*); and
also from my former *R. hirtus*, which is now thought to be
R. humifusus. It may be doubted if Dr Godron's plant is
identical with either my former or present *R. hirtus;* for

he mentions strong and wounding prickles amongst its characters.

It is highly probable that an examination of the many localities in central Wales, which closely resemble the lower parts of the valley of Llanberis, will show that this plant has an extensive range in that wooded country and damp climate.

Habitat.—Edges of woods in a damp climate. July, August.

Area.—1 . . . 5 . 7.

Localities.—i. Carclew, *E. Corn.;* near Plymouth, *S. Dev.;* Culbone, *S. Som.*—v. North side of Shrawley Wood, *Worc.*—vii. Abundant in the valley of Llanberis, *Caern.*

35. R. Güntheri Weihe.

R. caule arcuato-prostrato tereti, *aculeis tenuibus inæqualibus* e basi magna subcompressa declinatis, aciculis setis pilisque brevibus subæqualibus, foliis ternatis vel raro quinato-pedatis, *foliolis* inæqualiter vel duplicato-dentato-serratis subæqualibus, *planis* supra opacis pilosis *subtus* pilosis viridi-canescentibus vel subtomentosis, foliolo terminali obovato-acuminato, paniculæ angustæ foliosæ ramis distantibus ascendentibus racemosis paucifloris *rachide flexuosa*, aculeis tenuibus declinatis paucis, pilis et setis subæqualibus multis, sepalis ovato-lanceolatis a fructu laxe reflexis.

R. *Güntheri* Weihe in Bluff et Fingerh. Comp. Fl. Germ. i. 679 (1825). Rubi Germ. 65. t. 21. Bab.! in A. N. H. xix. 17 ; Man. ed. 2. 105 ; ed. 5. 107 ; ed. 6. 116. Blox.! in Kirby 41. Lees! in Steele 55 ; Malv. 51. Syme's Eng. Bot. iii. 188.

R. *cinerascens* Bor. Fl. Centre, ed. 3. ii. 197 (1857).

R. *pyramidalis* Quest.! in Billot. exsic. 2058 (sp.).

R. *glandulosus var. subracemosus* Bab.! in Blox. Fascic. (sp.).

Stem arcuate-prostrate, round or slightly angular. *Prickles* very slender, declining, from a long slightly compressed base. *Hairs* few ; *aciculi* rather more numerous ; *setæ* abundant ; all short, nearly equal. *Leaves* ternate or quinate-pedate. *Leaflets* large, flat, unequally and doubly dentate or serrate, pilose above, paler, hairy, and often with a fine coat of short ashy hairs beneath ; basal unequally obovate, acuminate; terminal obovate, acuminate: quinate leaves rare but found on the same stems as the ternate leaves, their leaflets obovate, cuspidate, terminal acuminate-cuspidate; petioles flattened above; stipules very slender.

Flowering shoot from reddish scales, wavy, hairy, with many short nearly equal aciculi and setæ, which do not exceed the many hairs. *Prickles* very slender, a little deflexed or much declining, from large compressed bases. *Leaves* ternate. *Leaflets* obovate-cuspidate, nearly equal, dull green above, paler and felted beneath; two or three uppermost floral leaves simple, ovate or cordate-ovate, often lobed. *Panicle* long, leafy; rachis wavy (forming an angle at each leaf), and as well as the branches and peduncles bearing an abundance of short nearly equal hairs and purple setæ; branches straight, racemose, ascending, few-flowered many; ultra-axillary top with short corymbose very few-flowered branches decreasing upwards into simple peduncles. *Sepals* ovate-lanceolate, acuminate, with a long leaflike point, reflexed, setose, aciculate, hairy, felted. *Petals* narrow, lanceolate, acute, pale pink. *Filaments* white. *Anthers* greenish. *Styles* pink at their base. *Primordial fruitstalk* rather longer than the sepals. *Nut* $\frac{1}{2}$-ovate; inner edge quite straight, except near the base where it projects in a remarkable manner to form the attachment to the receptacle; style quite lateral, seeming to tip the inner edge of nut. The nut is about as broad, but considerably longer than that of *R. pyramidalis*.

The armature of the stem of Mr Lees' plant from Crows-nest wood is much stronger than is usual, and greatly approaches that which is found in the *Koehleriani*. Some of the specimens distributed by him might well be placed in that group, others more exactly belong to the *Bellardiani*. Another proof, if one was wanting, of the very artificial character of our sections.

I believe this to be the *R. hirtus β Menkii* of Lees
(1847), and of his *Botany of Worcester*, from possessing a
specimen gathered, named and given to me by that botanist
in 1849 as *R. Menkii* (W. and N.). It was gathered in
Shrawley wood, Worcestershire.

It has been already remarked under *R. pyramidalis*
that some botanists have been inclined to believe that that
plant is not really distinct from *R. Güntheri*. It seems
to me that the differences between them are amply sufficient
to show a distinction of species, and I am glad to find that
Mr Boswell Syme is of that opinion (*Eng. Bot.* iii. 188).
Had the *R. pyramidalis* been founded upon a few specimens
preserved in Herbaria, its distinctness might have seemed
to be doubtful; but the plant has been carefully studied in
a living state, in the valley of Llanberis, where it is remarkably abundant, and luxuriant, and constant to its characters.
It also preserves it characters when raised from seed and
cultivated at Cambridge, but does not luxuriate in the dry
climate of that place, as in the damp one of North Wales.
Likewise, *R. Güntheri* always possesses the structure described above, wherever it has been found. Mr Lees, whose
opinion is of much value from the careful attention which
he has paid to *Rubi*, separated *R. pyramidalis* from its ally
in the year 1849, and conferred upon it the name of
R. longithyrsiger, as I learn from a specimen from him
which is preserved in my Herbarium. I cannot learn that
he ever published that name with or without a description
of the plant. He now (*Bot. of Worc.*) places my *R. pyramidalis* under his *R. Menkii* (not of W. and N.), and
separates the species from *R. Güntheri*. He also states that
his *R. thyrsiflorus* (*Steele's Handb.* 56) is "a very dilated state"
of this species. The straight rigid rachis, pyramidal panicle,
want of felt on the under side of the leaves, and strong
prickles of *R. pyramidalis*, seem to be markedly different

from the—wavy rachis forming an angle at each joining, distantly, irregularly, and shortly branched panicle, leaves with a thin coat of felt beneath, and much more slender prickles of *R. Güntheri:* also the nuts are exceedingly different in shape.

There hardly can be any doubt of the identity of our plant and the *R. Güntheri* of the *Rubi Germanici.* Arrhenius considered it to be a state of *R. glandulosus* produced in shade ; but we find that that species does not take this form under such circumstances : also its panicle and arma- ture are very different, and its leaves are considerably dissimilar. It has much in common with *R. hirtus*, with which Godron combines it.

The *R. Güntheri* of Questier in Billot's *Fl. exsic.* (No. 2057) appears to be *R. rosaceus* : my *R. Güntheri* has been sent by him as a form of *R. dumetorum.*

The wavy rachis, reflexed fruit-calyx, and usually 3-nate leaves seems to distinguish *R. Güntheri.*

Habitat.—In shady places. July, August.

Area.—1 . 3 . 5 . . 8 . 10 . . . . 30.

Localities.—i. Cothele, *E. Cornw.;* near Plymouth, *S. Dev.*—iii. High wood near Bramfield, Prae wood near St Albans, Easney Park wood, near Welwyn, Oxhey wood (Hind). *Herts. ;* Barach wood, Warley, *S. Essex* (Hind !) ; Trent wood, *Middles.* (Hind).—v. Crows-nest wood near Worcester and Shrawley wood, *Worc.;* Atherstone Outwood, *Warw.;* Lydney, *W. Glouc.*—viii. Ashby de la Zouch, and in Buddon wood, *Leic.* (Blox. !).—x. Above Gormire, *N. E. York.*

    xxx. By the road from Garvagh to **Kilrea** *Derry* (D. Moore !).

36. R. humifusus Weihe.

R. caule arcuato-prostrato subtereti, aculeis crebris tenuibus valde inæqualibus e basi longa compressa declinatis, *aciculis tenuissimis cum setis et pilis inæqualibus* crebris et patentibus, foliis quinato-pedatis, *foliolis argutè sed duplicato-patenti-dentatis* supra pilosis opacis *subtus* pilosis viridi-canescentibus micantibus, foliolo terminali obovato-oblongo subcuspidato, paniculæ latæ infernè foliosæ ramis corymbosis rachide subflexuosa, aculeis tenuibus declinatis paucis, pilis et setis inæqualibus crebris, sepalis ovatis pauci-aciculatis brevi-setosis a fructu laxe reflexis.

R. humifusus Weihe in Bluff et Fingerh. Compend. Fl. Germ. i. 685 (1825). Rubi Germ. 84. t. 35. Bab.! Syn. 31; Man. ed. 2. 10; ed. 6. 116.

R. hirtus, a Bab.! Syn. 29; Man. ed. 2. 104; ed. 3. 102; ed. 4. 104; Ann. Nat. Hist. ser. 2. ii. 39.

R. pallidus β foliosus Lees! in Steele 59.?

Stem terete or slightly angular, prostrate almost from its base. *Prickles* many, very unequal, very slender, declining or slightly deflexed from a long compressed base. The *hairs*, *setæ* and very slender *aciculi* many, unequal, patent, merging very gradually into each other as do the aciculi into the prickles. *Leaves* quinate-pedate. *Leaflets* finely but doubly dentate, dull green and pilose above, paler, hairy and often with a fine coat of short ashy hairs beneath; basal lanceolate; intermediate obovate subcuspidate, rather unequal below; terminal obovate-oblong, subcuspidate, cordate at the base; all stalked; rarely the basal and intermediate combine into a lobed leaflet; petioles (which are flat or slightly convex above) and midribs with small slender hooked or declining prickles beneath; stipules linear-lanceolate.

Flowering shoot armed like the stem but less strongly. *Leaves* ternate. Leaflets obovate cuspidate, often much narrowed below, dull green and pilose above, hairy and with whitish ashy felt beneath; uppermost leaves often simple, cordate-trilobed or ovate. *Panicle* with few distant short corymbose erect-patent branches (except the lowest which is sometimes subracemose); top leafless, racemose, with short ascending few-flowered corymbose branches with long stalked terminal flowers; rachis slightly wavy, hairy, with many shortish but unequal setæ, few slender aciculi, and few slender declining long-based prickles; uppermost branches and peduncles similarly armed but so much more hairy as even to seem felted. *Sepals* ovate, with a slender point, greenish, felted, hairy, with sunken setæ and rarely a few slender aciculi, loosely reflexed from the fruit. *Petals* oval, notched, narrowed below, large, white. Apparently the *primordial fruitstalk* is longer than the calyx. *Nut* obovate; inner edge nearly straight. In the *Rubi Germanici* the filaments are represented as whitish, anthers purple, and styles green.

The felt sometimes found on the under side of the leaves is very inconstant. It is nearly, if not quite, wanting on some of Leighton's specimens from Almond Park, whilst it is very apparent on others from the same place. The panicle is sometimes nearly simple; even the axillary branches being reduced to simple peduncles.

The figure and description of *R. humifusus* given in the *Rubi Germanici* agrees so well with our plant, that I have no doubt of their belonging to the same species. I refer my former *R. humifusus*, from Inverarnan, to this species. It is not the *R. hirtus* (Weihe), as figured and described in the above-mentioned work, which seems to be very nearly allied to *R. Bellardi*, and will be found, together with that plant, placed under *R. glandulosus* in this essay.

Mr Lees' *R. pallidus β foliosus* is apparently a form of this species. My specimens of it were gathered at the foot of the Great Doward Hill, Herefordshire. Its stem is much more prickly; its panicle very long, leafy near the top, with large and mostly simple leaves. I suspect that the plant was in an unnatural condition, caused probably by peculiarity of situation. The panicle of my specimen was gathered when too young.

Mr Baker finds an interesting plant on the hill side above Byland Abbey in Yorkshire, which I refer to this species; but M. Genevier says it is very closely allied to *R. offensus* (Müll.) of which I cannot find any account; it has ternate leaves and thus confirms my opinion that *R. humifusus* is one of the *Bellardiani*. Mr Baker considers it to differ from *R. humifusus* by its "more hairy leaves and adpressed sepals." I do not consider the former difference of much consequence, indeed some of my specimens of *R. humifusus* have quite as hairy leaves; and the sepals are loosely adpressed. He finds another plant near the same place, which M. Genevier names *R. saxicolus* Müll., and which I consider as hardly differing from our *R. humifusus* in any respect.

I now refer the plant mentioned in my *Synopsis* and in *Flora Hertfordiensis* as *R. horridissimus* to the present species.

Garke and Sonder combine *R. humifusus* with *R. pygmæus* (Günth.). I have no practical acquaintance with that plant, but Metsch states that he has examined the original specimens of Weihe, and finds that *R. humifusus* is certainly

21

not the same as *R. pygmæus*.   He also says that *R. humifusus* probably forms one species with *R. Schleicheri* (Weihe). Leighton's *R. Schleicheri* is quite a different plant.

Habitat.—Woods and thickets.   July, August.

Area.—. . 3 . 5 . 7 . 9 10 11 12 13 . 15 . . . .
. . . . 30.

Localities.—iii. Easney Park wood, *Herts.;* Messing wood, *N. Essex* (Varenne).—v. Bank of Wye below Great Doward Hill, *Heref.;* Almond Park, *Salop.*—vii. Craig Breidden, *Montg.*—ix. Beeston Castle, *Ches.* (Borr.!).*— x. Byland, *N. E. York.*—xi. Scots wood Dene, *Northumb.*— xii. Serbergham, *Cumb.* (Borr.!).

xiii. Jardine Hall, *Dumf.*—xv. Inverarnan at the head of Loch Lomond, *W. Perth.*

xxx. By the river Foyle above Londonderry, *Derry* (D. Moore!).

* Dr Bell Salter thought that this plant, of which a specimen will be found in the Herb. Borr. at Kew, is the *R. apiculatus* of Weihe.

37. R. foliosus Weihe.

R. caule arcuato-prostrato angulato, aculeis crebris tenuibus inæqualibus e basi longa compressa declinatis, aciculis tenuissimis æque ac setis sparsis inæqualibus, pilis paucis, foliis quinato-pedatis, *foliolis inæqualiter dentato-serratis* supra pilosis opacis *subtus pallidioribus pilosis*, foliolo terminali rotundo-cordato-acuminato, paniculæ longæ angustæ ad apicem foliosæ ramis brevibus corymbosis erecto-patentibus rachide subflexuosa, aculeis tenuissimis declinatis crebris, pilis et setis inæqualibus crebris, *sepalis ovato-attenuatis aciculatis setosis* hirtis a fructu laxe reflexis.

R. foliosus Weihe in Bluff et Fingerh. Compend. Fl. Germ. i. 682 (1825). Rubi Germ. 74. t. 28. Bab.! Man. ed. 5. 108; ed. 6. 117.

R. hirtus γ *foliosus* Bab.! in A. N. H. Ser. 2. ii. 39; Man. ed. 3. 102 ; ed. 4. 105.

R. exsecatus Müll.! in Wirtg. Herb. Rub. No. 179 (sp.) (1862).

Stem arcuate-prostrate, slightly angular, with many prickles aciculi setæ and hairs at its base ; rather more angular above, but with fewer setæ and aciculi, which merge gradually into slender, declining, rather long-based prickles, hairs few. *Leaves* quinate-pedate. *Leaflets* all stalked, unequally apiculate-dentate, with the teeth rather pointing forwards, slightly wavy at the edge, slightly pilose, dark green and opaque above, rather paler and pilose on the veins beneath; basal ovate-acuminate ; intermediate obovate-acuminate, unequal and sometimes subcordate at the base ; terminal roundly cordate acuminate ; petioles (which are slightly furrowed above) and midribs with rather many slender, declining or deflexed prickles beneath ; stipules linear.

Flowering shoot with few small slender long-based declining prickles, aciculate, setose, hairy. *Leaves* ternate. *Leaflets* coarsely doubly serrate, slightly pilose dark green and opaque above, rather paler and hairy on the veins beneath ; basal shortly stalked, unequally ovate, acute ; terminal broadly ovate-acuminate. *Panicle* with many long slender long-based declining prickles, and many very unequal aciculi and setæ, very long, narrow, leafy to the top ; uppermost floral leaves simple ; lowest branches racemose ; others corymbose, about 3-flowered, with peduncles of nearly equal length ; branches and peduncles very hairy setose and aciculate ; rachis slightly wavy, less densely clothed than the peduncles. *Sepals* ovate-attenuate with a slender point, very hairy, bearing many long setæ and aciculi, loosely reflexed from the fruit. *Petals* obovate, clawed, distant, white, slightly crenate at the end. *Filaments* white. *Anthers* yellowish. *Styles* greenish. *Primordial fruit-stalk* rather short. *Nut* half-ovate ; inner edge straight.

Mr Bloxam remarked in 1847 that "this plant seems to accord exactly with the figure and description in the *Rubi Germanici.*" But there are some slight differences between our specimen and that described and figured in that work : chiefly that the terminal leaflet of our plant is much more markedly cordate below ; the panicle-branches are more patent, and the calyx is differently clothed. In all these respects it agrees much better with the *R. exsecatus* (Müll.), with which indeed I think it is identical. Weihe and Nees remark that the sepals of their plant are "præter tomentum neque glanduli neque alio armorum genere instructi :" very different from the extremely hairy and abundantly setose sepals of our plant and of *R. exsecatus;* in the latter the hair is rather less abundant, and the setæ are more numerous.

This plant is allied to *R. humifusus:* but the serration of the leaves is peculiar, and the ovate-attenuate spinous sepals are very unlike those of that species. The shape of the terminal leaflet also is remarkable.

Should it ultimately be determined that this is not the *R. foliosus* (Weihe), as seems quite possible, the name given by Müller must probably be adopted for it.

I believe that Mr Bloxam has not published any account of this plant, nor do I know if it is abundant at Hartshill wood, from whence I possess two excellent specimens, or at the other station near Atherstone. Mr Briggs tells me that it is found in several places near Plymouth.

Habitat.—Heaths and woods. July, August.

Area.—1 . . . 5.

Localities.—i. Plymouth, *S. Dev.* (Briggs!).—v. Hartshill wood and (according to Mr Syme) Annesley Coal-field Heath, both near Atherstone, *Warw.*

38. R. glandulosus Bell.

R. caule arcuato-prostrato subtereti, aculeis parvis è basi longa compressa declinatis, aciculis setis pilisque subæqualibus crebris, foliis ternatis vel raro quinatis, foliolis subæqualibus oblongis cuspidatis subtus in venis tantum pilosis, foliolo terminali subcordato-ovato-acuminato, paniculæ tomentosæ valde setosæ aciculatæ ramis erecto-patentibus axillaribus apice racemoso, aculeis tenuibus declinatis, sepalis ovato-attenuatis aciculatis setosis tomentosis fructui laxe adpressis vel patentibus.

R. glandulosus Bellardi in Mem. Acad. Turin. v. 230 (1791). Tratin. Ros. iii. 21. Poir. Encycl. Method. Suppl. iv. 694. Bab.! Man. ed. 5. 108 ; ed. 6. 117, Syme's Eng. Bot. iii. 190.

R. hybridus Wallr. Sched. 229. Vill. Pl. de Dauph. iii. 559 (1789) (?). Garke Fl. v. Nord und Mitt. Deutschl. ed. 125. 7.

a.   Bellardi; foliis ternatis, foliolis subtus in venis brevi-pilosis argute dentato-serratis subæqualibus oblongis lateralibus divaricatis, paniculæ ramis paucis axillaribus distantibus corymbosis rachide sæpe flexuosa.

R. Bellardi Weihe in Bluff et Fingerh. Compend. Fl. Germ. i. 688 (1825). Rubi Germ. 97. t. 44. Wimm. et Grab. Fl. Siles. ii. 41. Lees! in Steele 55 ; Bot. Malv. 41. Wimm. Fl. Schles. 134. Billot.! Fl. Gall. et Germ. exsic. No. 1869 (sp.).

R. hirtus Reichenb. Fl. excurs. 607 (1830).

R. glandulosus Borr.! in Eng. Bot. Suppl. 2883. Arrh.! Mon. 40. Fries! Nov. Mant. altera. 36 ; Summa 167 ;

Herb. Norm. v. 52 (sp.). Metsch in Linnæa xxviii. 175.
Billot! Fl. Gal. et Germ. exsic. No. 2257 (sp.). Syme's
Eng. Bot. iii. t. 454.

R. glandulosus a *Bellardi* Bab.! Man. ed. 6. 117.

R. Wirtgeni Auersw. in Wirtg. Fl. Preuss. Rhein. 155.

Stem arcuate-prostrate, round below, slightly angular
towards the extremity, dark red when exposed, with a slight
glaucous bloom, densely covered with short and nearly equal
red aciculi and setæ, and a few hairs. *Prickles* short and
slender, from longitudinally dilated bases, all nearly equal and
longer than the aciculi, declining. *Leaves* ternate. *Leaflets*
nearly equal, large, convex, oblong, cuspidate, green on both
sides, finely dentate-serrate, pilose above, with short hairs on
the veins above, and slightly paler beneath ; basal unequal-
sided, patent, shortly stalked ; terminal slightly obovate,
stalked, rounded or subcordate below. Rarely the leaves are
quinate ; lower leaflets oblong, cuspidate, shortly stalked ;
intermediate obovate-oblong, cuspidate, rather unequal-sided
and subcordate at the base, stalked ; terminal obovate-oblong,
cuspidate, cordate at the base. *Petioles* (which are flat above)
and midribs with many small unequal declining or deflexed
prickles, aciculi and setæ beneath ; stipules very narrow or
linear-lanceolate-attenuate.

Flowering shoot from fuscous scales, armed like the stem.
Leaves ternate. *Leaflets* obovate, narrowed below ; basal
unequal-sided ; or rarely the leaves are simple, 3-lobed, with
a cordate base. *Panicle* broad and short ; branches few,
straight, axillary, short, patent, usually corymbose, with the
lateral flowers patent and long-stalked, and the terminal
flower shortly stalked, or racemose-corymbose, or the lowest
racemose ; top racemose, short ; prickles very slender ; setæ,
and hairs many ; rachis rather wavy, sometimes remarkably
so. *Sepals* ovate-attenuate, leaf-pointed, aciculate, setose,
hairy, felted, adpressed to the young fruit, afterwards more

or less reflexed. *Petals* distant, narrow, obovate, narrowed below, entire, white. *Filaments* white. *Anthers* and *styles* greenish, the latter sometimes rather pink at their base. *Primordial fruit-stalk* short, about as long as the sepals. *Nut* ½-ovate ; inner edge straight.

Perhaps the most remarkable points observable in this plant are the ternate leaves with the lateral leaflets placed opposite to each other and at right angles to the pedicel of the terminal leaflet; and the very open but short and usually wavy panicle with the axillary branches spreading nearly at right angles, undivided below, and ending in a simple or double corymb of flowers ; the branch is quite straight from its base to the terminal flower, so are the secondary branches, which are themselves patent, and bear patent lateral flowers. The terminal flower of each branch has usually a shorter stalk than the lateral flowers.

This plant sometimes has quinate leaves, and is much stronger than when they are ternate ; its leaflets are very much larger, as also is its panicle. It does not seem to differ in other respects, and is then apparently a plant of woodland districts, and shows an approach to the *var. hirtus.* Mr Bloxam identifies the plant found at Terrington Car with *R. Wirtgeni* (Auersw.) of Wirtgen's *Herb. Rub.*

Arrhenius has proved that this is the *R. glandulosus* of Bellardi, in opposition to the opinon of Reichenbach and others that *R. hirtus* (W. and N.) is the true plant. His opinion is apparently fully confirmed by the remark of Bellardi in the original description of his plant—"folia in meis speciminibus nunquam quinta;" therefore the conclusion deduced by Reichenbach from the expression "foliis quinatis ternatisque" appearing in Bellardi's specific character of the species must be rejected. Wahlenberg, who saw Kitaibel's specimens, describes our plant as *R. glandulosus*, and adds, "Ab hoc distinguitur *R. hirtus* (Kitaib.)

caule angulato, foliis subtus pilis splendentibus fere incanis, aculeis in caule petiolo paniculâque duris compressis et basi decurrentibus, hirsutie in petiolis pedunculis et calycibus albonitente eglandulosa, in qua raro una alterave glandula detegitur. Color foliorum saturate viridis." *Wahl. Fl. Carpat.* 152.

The specimen given by Reichenbach in my copy of his *Flora exsiccata* (No. 875) as *R. glandulosus* (Bell) is very young and incomplete, being only the top of a flowering shoot. Its rachis, peduncles, and especially calyx, are covered with very many, exceedingly long, purple setæ, intermixed on the rachis and peduncles with an abundance of very slender declining prickles, which merge gradually into aciculi, and these latter into setæ. The rachis is wavy. The lower branches racemose; the upper subcorymbose. It does not agree with the plates or descriptions of either *R. glandulosus* or *R. hirtus*, and is not quoted under either of those names by Godron.

Sub-var. *dentatus;* caule subangulato, foliis ternatis subæqualibus ovatis acuminato-cuspidatis basi cordatis subtus cinereo-viridibus lateralibus patentibus ascendentibusque, reliquis ut in *R. Bellardi.*

R. dentatus Blox.! in Kirby 39 (1850).

R. glandulosus δ *dentatus* Bab.! in A. N. H. xix. 17; Man. ed. 2. 105.

R. Mülleri Wirtg. Herb. Rub. (teste Blox.).

This plant is so like the *R. Bellardi* that a full description is unnecessary. I see no reason to doubt their specific identity.

M. Questier has sent a specimen, which appears to be *R. dentatus* (Blox.), with the remark, "*R.* (olim mihi *Schleicheri*) nunc *Güntheri* forma, floribus quamvis roseis." It agrees exceedingly well with my specimen of *R. dentatus*

(Blox.), with the exception of the terminal leaflet, which is more ovate and more cuspidate.

β. *hirtus;* foliis quinatis, foliolis subtus in venis longe et dense pilosis micantibus grosse inæqualiter serratis, foliolo terminali subcordato-ovato-acuminato, paniculæ sæpe elongatæ ramis racemosis vel corymbosis brevi-setosis, rachide subrecto.

R. hirtus Wald. and Kit. Pl. Hungar. ii. 150. t. 141 (1805). Weihe in Bluff et Fingerh. Comp. Fl. Germ. i. 688. Rubi Germ. 95. t. 43. Trattin. Ros. iii. 23. Wimm. et Grab. Fl. Silec. ii. 38. Wimm. Fl. Schles. 134.

R. hirtus a *Weiheanus* Metsch in Linnæa xxviii. 160.

R. glandulosus β *hirtus* Bab.! Man. ed. 5. 108; ed. 6. 117.

R. glandulosus Reichenb. Fl. excurs. 607.

R. fuscus Lees! in Steele 55 (1847); Malv. 52. Blox.! in Kirby 40 (not of Weihe).

R. glandulosus β *fuscus* Bab.! Man. ed. 4. 105.

R. glandulosus γ *rosaceus* Lees! in Steele 55.

R. fusco-ater Lindl.! Syn. ed. 1. 94. Leight.! Fl. Shrop. 235.

R. Koehleri ε *fuscus* Leight.! Fasc. 21 (sp.).

Stem arcuate-prostrate, round at the base, angular above, with many hairs, much branched. *Aciculi* and *setæ* unequal, few, short. *Prickles* many, rather strong but slender, much declining from a long compressed base. *Leaves* quinate-pedate. *Leaflets* all stalked, green on both sides, pilose above, with many long shining hairs on the veins beneath, coarsely and irregularly serrate; basal oblong; intermediate, obovate, acuminate; terminal ovate, acuminate, subcordate below; petioles and midribs with many hooked prickles beneath; stipules linear.

Flowering shoot from reddish brown scales, very hairy,

setose, aciculate. *Prickes* very small, slender, declining. *Leaves* ternate. *Leaflets* rather coarsely and doubly serrate. *Panicle* long; branches ascending, long, axillary, racemose or corymbose; top leafless, racemose, with divaricate few-flowered branches having the stalk of their terminal flower usually about as long as the ascending stalks of the lateral flowers which are nearest to it; rachis more or less wavy. *Sepals* ovate-attenuate, with a leaflike point, hairy, setose, aciculate, clasping the fruit. *Petals* distant, roundish, blunt, entire, clawed, white. *Filaments* white. *Anthers* greenish. *Styles* faintly flesh-coloured. *Primordial fruitstalk* short, equalling the calyx. *Nut* $\frac{1}{2}$-ovate; inner edge nearly straight.

Garke (l. c. 124) considers the *R. hirtus* (W. and N.) as distinct from that of W. and K., and combines *R. Güntheri* with it. I have no doubt that the present plant is the *R. hirtus* (W. and K.) and probably also of Weihe and Nees.

My specimens vary considerably in the amount of hair upon the leaves, especially on their upper side; but this seems to result rather from its being deciduous than originally wanting there. When the panicle is well developed it corresponds with the plate in *Rubi Germanici*, but it is frequently very much smaller and less branched. Mr Lees remarks that the panicle is often like that of *R. thyrsiflorus* (Weihe), and as that closely resembles the same part in *R. hirtus* (judging from the plates) his opinion may be considered as corresponding very accurately with mine. Unfortunately I had misled him and others into the idea that our *R. humifusus* was the *R. hirtus*, and thus he was prevented from expecting to find his *R. fuscus* under that name. Mr Lees' *R. hirtus* is shown by an authentic specimen to be *R. fusco-ater*. The plant given by Mr Bloxam (*Fasciculus*) as *R. fuscus*, because so named by Mr Lees, is nevertheless not the latter botanist's plant, and may be *R. Radula*. It is from Great Cowleigh Park.

I was long inclined to consider *R. hirtus* as distinct from
R. Bellardi; but the examination of an extensive series of
specimens gathered at Seckley wood (a part of Wyre Forest)
has convinced me that they are forms of one species.

The plant named *R. fusco-ater* by Lindley for Leighton
seems to be *R. hirtus;* that of the Hort. Soc. Garden was
R. Koehleri γ pallidus.

Sub-var. *rotundifolius;* caule subangulato, foliis ternatis, vel raro quinatis, foliolis duplicato-dentatis cuspidatis terminali subrotundo basi subcordato, reliquis
ut in *R. hirto.*

R. rotundifolius Blox.! in Kirby 39 (1850).

R. glandulosus ϵ *rotundifolius* Bab.! in A. N. H. Ser. 2.
ii. 40.

R. Lejeunii Lees! Malv. 52.

R. glandulosus β *Lejeunii* Lees! in Steele 55.

If the broadness of the leaflets, especially the terminal
one, and the greater regularity of the toothing are not considered, this plant agrees admirably with the *R. hirtus*
(Weihe). It seems to be almost certainly a form of that
plant with leaves which are usually ternate and have the
lateral leaflets divaricate and very gibbous or lobed on the
lower edge. A full description does not seem requisite. A
remark in the *Flora of Leicestershire* might convey the idea
that I formerly considered this plant as *R. rosaceus* (Weihe),
but Mr Bloxam informs me that that was not his intention.

Habitat.—Woods. July, August.

Area.— . 2 3 . 5 . 7 8 . 10 . . . . 19.

Localities of *a.*—v. Near Tintern, *W. Glouc.;* Chase wood,
near Ross, *Heref.;* Cowleigh Park, *Worc.;* Seckley wood,
Salop.—vii. Llanberis, *Caern.*—x. Terrington Car, *N. E.
York.*

xix. Foot of Turk Mountain at Killarney, *S. Kerry.*

Localities of *dentatus.*—v. Atherstone, *Warw.* (Blox.).— viii. In a fir plantation and in hedges by the Appleby road, near Twycross, *Leic.*—x. Loxley near Sheffield, *S. W. York.*

Localities of *β.*—ii. Wakehurst, *W. Suss.* (Mitten!)—iii. Welwyn road, Panshanger, *Herts.*—v. *Monmouth;* Western base of Malvern hills and near Ross, *Heref.;* Gt. Malvern and Gt. Cowleigh park, *Worc.;* Foot of Wrekin, *Salop.*— vii. Rhayader Mawddoch, *Merion.* (Borr.!); Lydney, *W. Glouc.* —viii. Twycross, *Leic.*—x. Bilsdale, *N. E. York.*—Guernsey.

Localities of *rotundifolius.*—v. Cowleigh park, *Worc.*— viii. In a fir plantation by the Appleby road, near Twycross, *Leic.*—x. Loxley near Sheffield, *S. W. York.*

Group V. Cæsii.

Caules sæpissime arcuato-prostrati, teretes vel sub-angulati, pruinosi. Aculei inæquales. Aciculi setæ pilique pauci vel nulli.

This is a very natural group of plants, and for that reason very difficult to divide into its true species. The limits of several of those which I have adopted are far from being known with certainty.

Although these plants clearly belong to the glandular section of the genus, the setæ and aciculi are often very far from abundant. They are usually difficult to detect upon the stems of *R. corylifolius* and *R. Balfourianus*, and those plants might easily be supposed to form parts of the group *Villicaules*, if they did not possess scattered and rather unequal prickles and more or less pruinose stems.

39. R. Balfourianus Blox.

R. caule arcuato-prostrato terctiusculo patenti-piloso, aciculis sctisque paucis, aculeis tenuibus in-æqualibus sparsis è basi oblonga subcompressa patentibus, foliis quinatis, *foliolis* dentato-serratis utrinque viridibus supra pilosis rugosis *subtus hirtis* nec tomentosis, foliolo terminali cordato vel ovato acuto infimis subsessilibus intermediis incumbentibus, paniculæ laxæ foliosæ hirtæ pauci-setosæ ramis longis distantibus pauci-floris racemoso-corymbosis erecto-patentibus, *sepalis ovato-acuminatis erecto-patentibus*, stylis dilute carneis, fructu hemispherico, toro oblongo pedicellato.

R. Balfourianus Blox.! in Fascic. of Rubi. 1846 (sp. and name only). Bab.! in A. N. H. xix. 86 (1847); Man. ed. 2. 100; ed. 5. 108; ed. 6. 118. Billot! Fl. Gall. et Germ. exsic. No. 1471 (sp.). Syme's Eng. Bot. iii. 192.

R. fusco-ater δ *subglaber* Bab.! in A. N. H. xix. 87 ; Man. ed. 2. 104.

R. tenuiarmatus Lees! Malv. 51 (1852).

R. Salteri β *Balfourianus* Bell Salt. in Bot. Gaz. ii. 120; in Hook. and Arm. Br. Fl. ed. 7. 125.

R. Schleicheri Lees! in Steele 54.

R. vulgaris Lindl.! Syn. ed. 2. 93.

R. corylifolius Johnst. E. Bord. 62.

Stem arcuate-prostrate, round near the base, angular towards the end, much branched, hairy with scattered patent hairs, a few subsessile glands, a few (rarely many) short equal setæ, an occasional aciculus, and sometimes a glaucous bloom. *Prickles* chiefly on the angles, slender, unequal,

patent, from an oblong rather cushionlike base. *Leaves*
quinate. *Leaflets* large, dull green rugose and pilose above,
paler and often so densely covered with silky hairs beneath
as to seem felted, although the actual surface is glabrous,
dentate-serrate in a rather irregular manner or sometimes
doubly serrate; basal subsessile, oblong or obovate, unequal
sided below, usually overlapping the intermediate leaflets
which are broadly oval or lanceolate; terminal roundly
cordate-cuspidate or oval-acuminate on the same plant:
rarely the leaves are ternate with very large leaflets of
which the basal are strongly lobed on the outer edge, and
the terminal is sometimes cordate-sub-3-lobed; petioles
(which are furrowed above) and midribs with small slightly
declining prickles beneath; stipules narrowly lanceolate.

Flowering shoot from reddish-brown scales, round,
clothed with woolly hairs or sometimes nearly glabrous,
with rarely a few aciculi and setæ, prickly like the stem.
Leaves ternate. *Leaflets* clothed like those of the stem,
coarsely or doubly dentate; basal sessile, strongly lobed
externally; terminal broadly obovate or lanceolate. *Panicle*
corymbose; having many patent hairs, rather few setæ,
very few aciculi; branches corymbose although few-flowered,
or simple and 1-flowered, erect-patent; one branch (usually
the lowest and the only one that is axillary) resembling and
often nearly equalling the rest of the panicle; sometimes the
branches are very long and very few-flowered so as to form
an exceedingly diffuse panicle; peduncles more hairy and
setose, felted; rachis wavy, seeming to be forked where
the main branch is given off. *Buds* depressed. *Sepals*
ovate-acuminate, greenish, hairy, shortly-setose, felted, often
leaf-pointed, erect-patent. *Petals* contiguous, roundly oval,
denticulate, pale pink. *Filaments* pale pink. *Anthers*
greenish. Styles flesh-coloured. *Fruit* oblong, sometimes
very large, black purple, having a slight taste of mulberry:

torus oblong ; *primordial fruit-stalk* scarcely ever as long as the sepals which are patent but bend upwards so as to clasp the fruit. *Nut* very broad, roundly $\frac{1}{2}$-ovate ; inner edge straight or concave.

The original *R. Balfourianus* is usually an exceedingly luxuriant plant, with enormous leaves upon both shoots, and a very large and very loose panicle. The first step from this is my former *R. fusco-ater δ subglaber* which has a large diffuse, much more prickly, but less hairy, although finely felted panicle. It has more and stronger but short aciculi on the stem, and leaves with fewer and shorter hairs beneath. The next step is formed by a plant having a small diffuse and corymbose panicle. And, lastly, I am unable to separate from the preceding the *R. tenuiarmatus* (Lees) my authentic specimen of which has a long leafy panicle with a corymbose top, and rather short and slightly racemose branches. The weak and abundant prickles, upon which the name is founded, "broken at the slightest touch," shrink after the specimens are gathered so as to become exceedingly compressed, but seem to spring from a somewhat cushionlike base which is oblong but not compressed : on the older stems this tendency to shrink ceases, but the prickles are very slender and much compressed. They are accompanied by plenty of short and strong aciculi. Typical *R. Balfourianus* has very few aciculi and much fewer prickles than *R. tenuiarmatus.*

Although the typical forms are very different, it is not always easy to distinguish this plant from *R. corylifolius;* for the *R. tenuiarmatus* approaches it closely. Usually the hairy but not felted under side of the leaves, the open panicle with scattered flowers, together with the long much more hairy and often leaf-pointed erect-patent sepals, will separate the *R. Balfourianus* from its ally : but sometimes the under side of the leaves of *R. corylifolius* is scarcely

if at all felted ; in rare cases the panicle is similar to that
of the small forms of *R. Balfourianus;* and the sepals clasp
a small nearly abortive fruit. Although we may not now
know the true limits of the species it seems to me very
highly probable that they are distinct.

A plant which grew at Henfield, Sussex, in 1845, and was
named *R. nemorosus* by Borrer, was placed in my Herbarium
as *R. Balfourianus.* The specimens were gathered in October
and the stems are nearly as naked as those of *R. Balfourianus,* but may have become so by the aciculi and setæ
having fallen off. Others gathered in the month of July
of the preceding year, but apparently not at precisely the
same spot, are similar in all respects, except that some parts
of the stems are very fully clothed with those organs and
an abundance of very unequal prickles. These latter specimens show that the plant (on which there seems to have
been no bloom) belongs to the *Koehleriani* not the *Cæsii.*
I place it under *R. fusco-ater.*

I have a specimen from Mr Bloxam, which he gathered
at Twycross, and called *R. corylifolius* in 1846. It seems
to be *R. Balfourianus,* but has the terminal leaflet of one of its
leaves divided into three with the central segment stalked ;
in another that leaflet is undivided and cordate-prolonged.

Brambles to which Mr Hort gave the provisional name
of *R. multiceps* may perhaps be placed here, although they
may, as I believe that he still suspects, be really distinct
from *R. Balfourianus.* They have prickles on the nearly
naked stem which closely resemble those of *R. tuberculatus.*
Their panicle is almost exactly like that of the less luxuriant
forms of *R. Balfourianus.* The terminal leaflet is elliptic.
Mr Hort also gave the name of *R. multiceps* to a specimen
which I gathered at Caerleon in Monmouthshire, and to
another found by himself by the river below the town of
Monmouth, both of which have a cordate-ovate terminal

leaflet. On the whole I think it best to place these plants with *R. Balfourianus*, and not to attempt to separate them even as varieties until we are better acquainted with them.

I possess a specimen of the *R. cæsius* δ *nudatus* of Lees (in *Steele's Handb.* 54) obtained from Leighton's Herbarium, to which it was given by Mr Lees. It grew at Henwick near Worcester, and very much resembles some states of *R. corylifolius* γ *purpureus*, but seems to possess the characters of *R. Balfourianus*.

M. Questier sent the *R. tenuiarmatus* (Lees) with the name of *R. Balfourianus*. His specimens agree well with the authentic plant of Lees. I also place here some specimens called *R. nemorosus* by M. Questier and myself. A specimen received as *R. dumetorum* from Mr Lange, gathered at Apenrade in Sleswig, is exactly the *R. Balfourianus*.

The above remarks will show that this is a very variable species which may ultimately require such division : but the series seems complete from typical *R. Balfourianus* to typical *R. tenuiarmatus*.

I am quite unable to conjecture the reasons which caused Dr Bell Salter to join this plant to *R. Salteri*, with which it seems to have very little in common.

Mr Borrer obtained a specimen of this species from the Horticultural Society's Garden as the *R. vulgaris* of Lindley who quotes *R. corylifolius* (Sm.) as a synonym in his second edition of the *Synopsis*. The *R. vulgaris* of his first edition is *R. villicaulis*.

I refer the *R. corylifolius* of Johnston (*E. Bord.* 62 and fig.) to *R. Balfourianus* with some slight doubt. He held the opinion that his *R. corylifolius* " is apparently different " from my plant so-named. His specimens and description agree in most respects with *R. Balfourianus*. As the specimens are in flower it is not possible to determine from them the condition of the fruit-calyx, and it is only on the living plant that the stamens and styles can be examined satisfactorily. I call the styles "flesh-coloured;" Dr Johnston said "yellowish-green changing to pink and brown," which differs perhaps more in appearance than reality from my terminology. Mr Bloxam's plant from Warwickshire "always grows in shaded hedges" and is "averse to the sun," which may account for its variation from the more usual forms of this very variable species.

It is possible that the *R. tiliæfolius* of Weihe (in *Spr. Syst.* ii. 529 and *D. C. Prod.* 562), published in 1825, may be this species, and if so, Bloxam's name would fall. A foreign specimen, given with that name but no locality to Mr Borrer by Mr Woods, is only the top of a panicle, but seems, as far as we can judge from such imperfect materials, to be *R. Balfourianus.* Without further and more conclusive evidence of their identity, we should not be justified in combining our plant with the *R. tiliæfolius* which Reichenbach tells us is his *R. corylifolius β pilosus* (*Fl. Excurs.* 608) and the *R. hirsutus* of Presl. (*Del. Prag.* 221). It is also the *R. dumetorum β pilosus* of the *Rubi Germanici*, 99.

The *R. magnificus* (Müll.!) of which I have not seen any description, is very like our *R. Balfourianus*, if not identical with it. Genevier states that it is the *R. Lejeunii* (Gen. et Godr.), and the *R. Bloxamii* (Bor.). Specimens from Yorkshire named *R. rivalis* by M. Genevier I also place here.

Habitat.—Hedges. July, August.

Area.—1 . 3 4 5 . . . 9 10 . 12 . 14 . . . . . . 21.

Localities.—i. Kingston, *S. Dev.* (Briggs!); by the canal at Claverton, *N. Som.*—iii. Mangrove Lane and Essendon, *Herts. ;* Red Hill and Capel (Borr.!), between Ditton Marsh and Claygate, *Surr. ;* Sheen, *Berks* (Bicheno); Woodend, *Middl.* (Hind); Tonbridge Wells, *W. Kent.*—

iv. Wicken Fen, Chesterton, and Toft, *Camb.*—v. Wyck and Stapleton, *W. Glouc.*; Chepstow, Newland, Ragland and Caerleon, *Monm.*; Bromsgrove Lickey, *Worc.*; Mill Lane, Coventry (Kirk), and Rugby, *Warw.*; Wistley Hill, Cheltenham, *E. Glouc.* (Notcutt).—ix. Rosthorne (Sidebottom in Bell Salt. Herb.!), Stretford (Hunt!), *Ches.*— x. Thirsk, *N. E. York.*—xii. Ambleside and Lowood, *Westm.*

xiv. Common in *Berw.* (Johns.!).

xxi. *Kilkenny.*

40. R. corylifolius Sm.

R. caule arcuato-prostrato teretiusculo vel obtus-
angulato subglabro, aciculis setis glandulisque raris-
simis, aculeis subulatis tenuibus subæqualibus e basi
longa subpatentibus vel raro deflexis, foliis quinatis,
foliolis duplicato-serratis utrinque viridibus supra sparsim pilosis rugosis *subtus* pallidioribus *tomentosis*, foliolo
terminali rotundo-cordato vel rotundo-ovato cuspidato
vel acuminato infimis subsessilibus intermediis incumbentibus, panicula ramisque subcorymbosis, *sepalis
ovatis cuspidatis a fructu reflexis*, petalis rotundo-
ovatis, stylis virescentibus, toro oblongo pedicellato.

R. *corylifolius* Sm.! Fl. Brit. 542 (1800); Eng. Fl. ii.
408.   Anders. in Trans. Linn. Soc. xi. 219.   Borr.! in
Hook. ed. 2. 248; ed. 3. 251.   Bab.! Man. ed. 1. 95; ed.
2. 106; ed. 5. 109; ed. 6. 118; Syn. 12; in A. N. H.
ser. 2. ii. 34.   Syme's Eng. Bot. iii. 192.

R. *affinis* Bab.! Man. ed. 1. 93.

a *sublustris;* caule teretiusculo rubro-viridi, *aculeis
tenuibus e basi oblonga* subpatentibus, foliolis subtus
cinereo-tomentosis *terminali sæpe subtrilobo* rotundo-
cordato, rachide teretiuscula pauci-aculeata.

R. *sublustris* Lees! in Steele 54 (1847); Malv. 51.
R. *corylifolius a sublustris* Leight.! in Phytol. iii. 161
(1848).   Bab.! Man. ed. 4. 106; ed. 6. 118.
R. *corylifolius* Sm.! Eng. Bot. t. 827 (1801).   Arrh.
Mon. 16.   Fries! Summa 168; Herb. Norm. vii. 48 (sp.).

Fl. Dan. t. 2538. Blox.! in Kirby 38. Syme's Eng. Bot. t. 455.

R. affinis γ Leight.! Fl. Shrop. 226.

R. nemorosus a glabratus Bab.! Syn. 32; Man. ed. 2. 106; ed. 4. 107.

R. maximus fructu nigro Linn. Wastgota Resa 135 (1747); Skanska Resa 139.

R. dumetorum "Auct. Helveticæ et præsertim Rapin Guide du Botan. dans le Canton de Vaud, 179." Genevier!

R. acerosus Müll.! (teste Genevier).

Stem arcuate-prostrate, terete or slightly angular towards the end, thick, very nearly or quite glabrous but with a very few scattered subsessile glands, setæ and aciculi, glaucous, and with scattered stellate down when young, usually greenish red. *Prickles* moderately abundant, rather unequal, slender, conical, slightly declining or patent with a longitudinally dilated but oval and usually small base. *Leaves* quinate. *Leaflets* nearly flat, wavy at the edge, doubly serrate, broad, dark green, slightly rugose, and with a few adpressed hairs above, paler, hairy, and felted (but sometimes only very finely) beneath; basal sessile, broadly oval, acute, overlapping the very broadly oval cuspidate intermediate leaflets; terminal roundly cordate with a small cusp, often having a large lobe on each side, and thus showing a tendency to divide into three; petioles (which are flat above) and midribs with short declining or deflexed prickles beneath; stipules broadly linear-lanceolate.

Flowering shoot from fuscous scales clothed with ashy silky hairs, roundish, slightly hairy. *Prickles* subpatent, large, from a long compressed base. *Leaves* mostly ternate, rarely quinate; uppermost usually simple, cordate-ovate or three-lobed. *Leaflets* whitish, hairy, and felted beneath; lateral ovate, unequal-sided; terminal obovate, broad, narrowed below. *Panicle* leafy below; branches corymbose,

ascending, axillary, long, naked, usually nearly or quite
without prickles towards their base; top corymbose, or with
a few short corymbose erect-patent ultra-axillary branches;
rachis nearly straight, and as well as the peduncles and
branches, felted and with small sunken setæ. *Sepals* ovate,
cuspidate, hairy, greenish, felted, with small sunken setæ,
reflexed from the fruit, but often closing over the remains of
an abortive flower. *Petals* contiguous, broad, roundly-ovate,
finely serrate, clawed, white ; or sometimes obovate and
narrowed below. *Filaments* white. *Anthers* greenish.
Styles yellow, but sometimes faintly pink at the base. *Primordial fruit-stalk* short, not as long as the sepals.

The true *R. sublustris* is exactly the typical *R. corylifolius*. It has a large, roundly-cordate, acuminate, more or
less 3-lobed terminal leaflet, which sometimes divides into
three distinct leaflets having the lateral sessile, and the
intermediate oval and shortly stalked. Owing to this tendency to divide, the leaflet is not quite constant in its form,
even upon the same bush, but its base is always cordate.
Sometimes the basal and intermediate leaflets on the same
side combine into a single bilobed leaflet.

There does not appear to be any doubt of the Swedish
R. corylifolius being identical with this variety to the exclusion of the others. It is also apparently the plant which
is carefully distinguished from *R. fruticosus* (*R. plicatus* or
R. discolor; probably the latter here) by Linnæus in his
Wastgota Resa, but unaccountably neglected in his systematic works. Richter (*Codex Bot. Linn.* No. 3760) considers that it was *R. cœsius* from which Linnæus distinguished it. He translates *Björnbär* by *R. cœsius*, but Linnæus in the *Flora Suecica* (ed. 2. 172) gives that Swedish
name to *R. fruticosus*. Arrhenius (p. 6) has a long note on
the subject, and considers the Linnæan *R. maximus fructu
nigro* to be *R. corylifolius* (Sm.).

β *conjungens;* caule subangulato rubro-viridi, *acu-leis* tenuibus validis *e basi longissima* compressa sub-patentibus sæpe apicibus paululum deflexis, foliolis subtus cinereo-tomentosis terminali cordato-ovato vel late obovato basi subcordato, rachide rectiuscula pauci-aculeata.

R. rhamnifolius Lind.! Syn. ed. 2. 92 (in part).

R. corylifolius β *conjungens* Bab.! Man. ed. 3. 103 (1851); ed. 5. 109; ed. 6. 118.

R. corylifolius β Leight.! in Phytol. iii. 161 (1848).

R. rhamnifolius (*second form*) Leight.! Fl. Shrop. 228 (in part).

R. sublustris γ *cœnosus* Lees! in Steele 54.

R. nemorosus γ *bifrons* Bab.! Syn. 32; Man. ed. 4. 107.

R. Wahlbergii Bell Salt.! (in part) in Ann. Nat. Hist xvi. 371; Fl. Vect. 159. Bab.! Syn. 31; Man. ed. 2. 106; ed. 3. 104 (in part).

Stem arcuate-prostrate, round at the base, with many small slender unequal prickles springing from roundish cushion-like bases, and many small setæ, angular with flat sides above, glabrous, slightly glaucous. *Prickles* nearly all upon the angles, short, rather strong, subpatent, from a long and compressed base, sometimes slightly deflexed at their tips. *Leaves* quinate. *Leaflets* glabrous and rugose above, whitish green hairy and felted beneath, nearly flat, doubly dentate; basal nearly sessile, ovate; intermediate shortly stalked, obovate, acuminate; terminal, shortly stalked, ovate or obovate, acuminate, more or less cordate below (sometimes very exactly cordate); petioles (which are slightly furrowed above) and midribs with few strong hooked prickles beneath;

23

stipules linear-lanceolate. Rarely a seta or aciculus may be found on the upper part of the stem.

Flowering shoot from brown silky scales, straightish, felted, especially towards the top. *Prickles* few, slender, declining, from large bases. *Leaves* ternate. *Leaflets* ovate, doubly dentate-serrate; those of the uppermost leaves pale green, felted, and hairy beneath. *Panicle* short, broad; top and branches subcorymbose; often consisting chiefly of two or three long axillary branches, themselves bearing terminal and lateral corymbs, and closely resembling (except in the rather looser arrangement of the flowers) the dense ultra-axillary top; rachis slightly wavy, and as well as the peduncles and branches felted, with a few short setæ, hairy. *Sepals* ovate, rather abruptly ending in a short linear point, hairy, felted, reflexed. *Petals* ovate-oblong, bluntish, slightly notched at the end, pink or white. *Filaments, anthers* and *styles* yellowish. *Primordial fruit-stalk* very short, shorter than the sepals. The panicle is sometimes leafy nearly or quite to its top.

I unfortunately once named a specimen of this plant *R. latifolius* for Mr Baker. Hence his erroneous idea of *R. latifolius* (*Suppl. to Baines's Fl. York.* 63. *Phytol.* iv. 969).

The usual form of this plant is described above, but a specimen before me deserves notice from the great difference which it presents. It has an enormously long panicle, leafy to its top, which is loosely corymbose with a long-stalked terminal flower ; the lower branches resemble the whole panicle on a small scale, but are leafless. This plant was gathered in Cambridgeshire by Mr Newbould, to whom I am indebted for the specimen.

The plant named *R. rhamnifolius forme ordinaire* by Nees v. Esenbech for Leighton, seems to be this variety of *R. corylifolius;* but it has scattered stellate pubescence upon its stem. The *R. nemorosus* γ *bifrons* of my *Synopsis,* the

R. corylifolius γ *cœnosus* (Lees), and the *R. Wahlbergii* of Salter and Bab. are forms of this variety with more and whiter felt on the leaves, and the *var. cœnosus* has an abundance of bloom on the stem. One of the plants from the Isle of Wight, doubtfully named *R. Wahlbergii* by Dr Salter and myself, is a slight variety of *R. althæifolius*, to which also a plant found near Henfield church, in Sussex, seems referable.

γ *purpureus;* caule angulato purpureo sæpe sparsim strigoso-sericeo, aculeis validis e basi longa compressa subpatentibus vel deflexis, foliolis subtus pallide viridi-albove-tomentosis terminali rotundo- vel subcordato-obovato, rachide subflexuosa multi-aculeata.

R. corylifolius γ *purpureus* Bab.! Man. ed. 3. 103 (1851); ed. 5. 109; ed. 6. 118.

R. corylifolius γ *Smithii* et δ *intermedius* Leight.! in Phytol. iii. 161 (1848).

R. corylifolius Leight.! Shrop. Rubi 6 (sp.).

R. rhamnifolius (second form) Leight.! Fl. Shrop. 228 (in part).

R. rhamnifolius Lindl. Syn. ed. 2. 92 (in part). Nees v. Esenb. in Leight. Fl. Shrop. 227.

R. Wahlbergii Arrh. Mon. 43. Fries! Herb. Norm. ix. 49 (sp.); Summa, 167.

R. nemorosus β *pilosus* Bab.! Syn. 32; Man. ed. 4. 107.

R. dumetorum α *glabratus* Lees! in Steele 54.

R. affinis γ Nees v. Esenbech in Leight. Fl. Shrop. 226.

R. thamnocharis Müll.! Mon. 190 (1859). Chab. Etude du Rubus, 30.

Stem arcuate-prostrate, round at the base, angular and often furrowed at the end, glabrous or thinly stellately downy, usually dark purple on the upper side, glaucous and with scattered stellate down when young; setæ and aciculi very few, except at the base. *Prickles* strong, nearly equal, abundant, slightly declining or slightly deflexed, from long compressed bases. *Leaves* quinate, a little concave. *Leaflets* flat, doubly dentate-serrate, dull green and pilose above, pale green or whitish felted and hairy beneath; basal sessile, obovate; intermediate broadly obovate, often unequal-sided, acute; terminal roundly obovate, often cordate at the base, cuspidate; petioles (which are flat above) and midribs with few strong hooked prickles beneath; stipules linear-lanceolate. Sometimes the basal and intermediate leaflets of each side combine to form single deeply lobed leaflets.

Flowering shoot from brown silky scales, slightly angular, glaucous, with scattered stellate down. *Prickles* rather abundant (especially towards the top of each internode in a more marked manner than in the other varieties), strong, slightly declining, from long compressed bases. *Leaves* mostly ternate; uppermost sometimes simple, three-lobed. *Leaflets* clothed like those of the stem; lateral ovate, unequal-sided, or lobed externally; terminal roundly obovate, cuspidate. *Panicle* leafy below, short; top and branches corymbose; often consisting only of the short broad naked top and two or three moderately long axillary branches; rachis often markedly wavy, but sometimes nearly straight; rachis branches and peduncles with a few short setæ, becoming more hoary with fine felt as the flowers are approached. *Sepals* ovate, rather cuspidate, hairy, felted, slightly setose, slightly aciculate towards the base, reflexed. *Petals* white or pink, roundly ovate, blunt, jagged, shortly clawed. *Filaments* purplish. *Anthers* yellow. *Styles* often pink at the base, otherwise greenish. *Primordial fruit-stalk* short, not so long as the sepals. *Nuts* unequally ovate; inner edge convex.

The specimen sent to Leighton by Lees as his *R. sublustris γ cænosus* differs from that which I received from him with the same name : the latter has a very pruinose stem without felt, and belongs to my *var. β conjungens;* Leighton's example has plenty of felt on its stem, and must be placed under my *var. γ purpureus.*

My acquaintance with the *R. Wahlbergii* (Arrh.) is limited to what can be derived from the single specimen contained in Fries' *Herbarium Normale* (ix. 49), which it may fairly be presumed is an authentic example of the plant. I am unable to distinguish this specimen from some forms of *R. corylifolius γ purpureus,* and do not think that it can be separated from this species. Upon a careful comparison of the Swedish plant with that variety, I find only the following slight differences :—The stem seems to be quite devoid of stellate down ; the stipules are much narrower than in *R. corylifolius;* but they are variable even in our plant; and I do not consider the presence or absence of diaphanous veins on them a satisfactory character for the distinction of species. Although the upper side of the leaves is described as "glaberrima," there are a few scattered hairs thereon in the Swedish specimen, which is in that respect precisely similar to some forms of our *R. corylifolius.* The colour of the filaments is white, whilst they are usually pink in my *var. γ purpureus:* they are more commonly, if not always, white in our other varieties. Although our plant has greenish styles they are occasionally tinged with pink at the base : a tint not mentioned as occurring in *R. Wahlbergii.* These are all the differences which I am able to detect by comparing the Friesian specimen with the above description of *R. corylifolius γ purpureus,* and similarly examining specimens of my plant with the description of *R. Wahlbergii* and the remarks published by Arrhenius. Certainly it is usual for that variety not to have deflexed

prickles, nor such very round and broad-based leaflets on the
flowering shoot; but the range of variation is very great
in both these respects. A specimen gathered many years
since near Bath, and named *R. corylifolius* by Mr Borrer, is
very exactly the *R. Wahlbergii:* others approach so closely
to this as not to admit of any doubt concerning their specific
identity with it. Metsch places *R. Wahlbergii* (Arrh.) as
a variety under *R. dumetorum* (Weihe). He seems to have
little acquaintance with the *R. corylifolius* (Sm.). Lange
(*Danske Flora*, 350) keeps it distinct, but erroneously refers
my *R. latifolius* to it.

The *R. pruinosus* (Arrh.) is exceedingly nearly allied to
R. corylifolius a sublustris. Of this, Arrhenius was well
aware. My knowledge of the species (if species it be) is
derived from the specimen contained in Fries' *Herb. Norm.*
(vii. 47), which was supplied to that valuable collection by
Arrhenius himself; from one authenticated by the same
botanist and kindly sent to me by Mr Lange, of Copenhagen ; and from another Swedish specimen, for which I am
indebted to Dr Lindeberg, of Göteburg. A careful comparison of these with *R. corylifolius* shows the following
difference. The stem is dark purple in colour, but is said
to be green in shady places (and is so in Dr Lindeberg's
specimen), with much more glaucous bloom than is usual
on our plant. The prickles resemble strong aciculi springing from oblong bases, and are slightly deflexed, corresponding exactly to those of some states of our *R. corylifolius*.
The prickles on Dr Lindeberg's specimen are considerably
stronger, but retain a similar general character. The terminal leaflets on the flowering shoot are very broad at the
base (except "near the lower end of the shoot"); our plant
seems always to have them narrowed below. The panicle
is leafy quite up to the top. In all other respects the *R.
pruinosus* seems identical with the smaller forms of *R.*

corylifolius a sublustris. Arrhenius states that the petioles of his *R. corylifolius* are furnished with straight prickles, whilst those of *R. pruinosus* are hooked : our *R. corylifolius* has them of both forms, even sometimes upon the same petiole. I think that our *R. corylifolius* is identical with that of Sweden, and I judge on this question also from the specimen communicated to Fries' *Herb. Norm.* (vii. 48), and Danish specimens from Mr Lange.

R. thamnocharis (Müll.) is probably identical with this variety, but approaches slightly to *var. β conjungens;* there I also place the *R. discoideus* (Müll.!) and *R. acanthophorus* (Müll.!).

Some remarks upon the plants now combined to form *R. corylifolius* will be found under *R. althæifolius* and *R. Balfourianus.* It is a variable species ; yet all its forms have a common look, which it is perhaps impossible to describe.

There can be no doubt that the plant intended by Smith was (typically) our *R. corylifolius,* although he probably included others under the name which are not now considered as correctly placed there. In the *Flora Britannica* he described the calyx as "maturascente fructus inflexus," on the authority of Mr Wigg ; but corrected the mistake in *English Botany* (f. 827).

M. Genevier refers specimens to the *R. Mougeoti* (Bill.).

There is a specimen in Billot's *Flora Gall. et Germ. exsiccata* (No. 763), which is named "*R. Wahlbergii* (Arrh., non Godr. Mon.) Gren. et Godr. Fl. Fr.*" and also stated to be the *R. dumetorum a vulgaris* of the *Rubi Germanici* (t. 45. A. f. 1). If placed by the side of that plate the difference between the plant and the figure will be seen to be very great. I am quite willing to believe that it is the *R. Wahlbergii* of the *Flora de France,* although the authors of that work quote the above figure from the *Rubi Germanici.*

If the specimen had any felt on its leaves I should refer it with certainty to *R. corylifolius;* but the underside of the leaves is very hairy on the veins and pale green in colour. Nevertheless, in my opinion, it is a feltless form of *R. corylifolius.* It seems to connect *R. tuberculatus* (Bab.) with *R. corylifolius.* As far as the single specimen will show, the barren stem is that of *R. tuberculatus,* but the panicle more resembles that of *R. corylifolius.* Can it be the *R. Holandrei* of Müller and of Chaboisseau (*Etude du Rub.* 29), to which the latter author refers the *R. Wahlbergii* of Godron and Boreau, the *R. plicatus* of Holandre?

Mr J. G. Baker, on the authority of specimens received from Wirtgen and Genevier, refers the following species of Müller to *R. corylifolius:* viz.

R. discriminatus,   R. permiscibilis,
R. malacophyllus,   R. ambiferius,
R. leucophæus,   R. dubiosus;

and also all the specimens received by him from Silesia as *R. dumetorum,* as well as the *R. dumetorum* of Wirtgen and the *R. patens* of Mercier.

Habitat.—Hedges and thickets.   June to August.

Area.—1 2 3 4 5 6 7 8 9 10 . . 13 . 15 16 . . . 20 . . . . . 26 . . . 30.

Localities of *a.*—ii. Albourne and Newtimber, *W. Suss.* (Borr.!); Hastings, *E. Suss.*—iii. Thames Ditton, *Surr.;* St Albans, *Herts.;* Harrow, *Middl.* (Hind).—iv. Fakenham, *W. Norf.;* Cambridge, *Camb.*; *Bedfordshire.*—v. Henwick, *Worc.;* Wellington and Shrewsbury, *Salop;* Lydney, *W. Glouc.*—vi. Milford Haven, *Pemb.*—viii. Twycross, *Leic.*—x. Settle, *N. W. York;* Thirsk, *N. E. York.*—xii. Douglas, *Isle of Man.*

xx. *Wicklow* (D. Moore!).—xxx. Funchanhale, Clondermot, and Templemore, *Derry* (D. Moore!); by Brett's Glen, *Down.*

Localities of β.—ii. Albourne, *W. Suss.;* Bembridge, *Isle of Wight, Hants* (Balfour !).—iii. Claygate, *Surr.;* Harrow and Notting Hill, *Middl.* (Hind); Purfleet, *S. Essex* (Sowerby!)—iv. Eversden, Grantchester, Madingley and Histon, *Cambr.;* Sandy, *Beds.*—v. Coleford, *W. Glouc.;* Castle Morton, *Worc.;* Shrewsbury, *Salop;* Winters Cross and Sellack Common, *Heref.* (Purchas!).—vi. Freshwater Bay east, *Pemb.;* Cardigan, *Card.*—vii. Bangor, *Caern.*—ix. Warrington, *S. Lanc.*—x. Thirsk, *N. E. York.*

xiii. Gouroch, *Renf.*—xv. Campsie, *Stirl.* (Hunt).—xvi. Lamlash, *Arran* (Balfour!).

Localities of γ.—i. Bath, *N. Som.*—ii. Henfield and Albourne, *W. Suss.*—iv. Fakenham, *W. Norf.;* Stetchworth, Kingston and Wisbech, *Cambr.*—v. Ross, *Heref.; Worcestershire;* Shrewsbury, *Salop.*—viii. Twycross, *Leic.*—x. Thirsk, *N. E. York.*

xvi. *Arran* (Balf.).

xxvi. Roundstone, *W. Galw.* (D. Oliver).—xxx. Brett's Glen, *Down.*

41. R. althæifolius Host.

R. caule prostrato subangulato sparsim piloso et setoso, aculeis crebris inæqualibus tenuibus è basi oblonga compressa patentibus, foliis quinatis vel ternatis, *foliolis crenato-lobatis* subtus pallide viridibus in venis pilosis vel laxe albo-tomentosis *inferioribus foliorum ternatorum retrorsum bipartitis quinatorum* sessilibus *intermediis dissitis, foliolo terminali rhombeo-obovato* basi subcordato, paniculæ foliosæ ramis axillaribus et apice racemoso-corymbosis setis paucis brevissimis, aculeis in medio rachidis quam reliquis longioribus tenuibus, *sepalis* ovato-subacuminatis setosis fructui (atro-cæruleo) *laxe adpressis*, petalis obovatis, stylis ad basin carneis.

R. althæifolius Host in Trattin. Ros. iii. 37 (1823); Fl. Aust. ii. 31. Ser. in DC. Prod. ii. 562. Bab.! Fl. Camb. 305 (1860); Man. ed. 5. 109; ed. 6. 119. Syme's Eng. Bot. iii. 193.

R. dumetorum Lindl.! Syn. ed. 2. 94.

R. dumetorum γ *tomentosus* Rubi Germ. 100. t. 45. A. fig. 2.   Metsch in Linnæa, xxviii. 115.

R. Wahlbergii Bell Salt.! in A. N. H. xvi. 371 (in part); Fl. Vect. 159 (in part).   Bab.! Man. ed. 3. 104 (in part).

R. Wahlbergii β *glabratus* Bell Salt.! in Fl. Vect. 160 (syn. excl.).

R. deltoideus P. J. Müll.! in Flora 181 (1858).

R. calcareus P. J. Müll.! in Flora 181 (1858).

R. virgultorum P. J. Müll.! Mon. Rub. 200 (1859).

Stem prostrate, round at the base, soon becoming bluntly angular, sometimes furrowed near the end, often nearly glabrous or with a few scattered hairs and very short deciduous setæ, green with a glaucous bloom or purplish. *Prickles* slender, many, unequal, almost setaceous near the base of the stem, slightly deflexed on the autumnal shoots, elsewhere conical, patent, springing from an oblong cushion-like base. *Leaves* quinate or ternate. *Leaflets* thin, flat, rugose, crenate-lobate-serrate, green on both sides, slightly pilose above, rather paler and slightly hairy on the veins beneath, or rarely densely hairy and felted; basal oblong, sessile, or extremely shortly stalked; intermediate obovate-acuminate, unequal-sided below; terminal broadly obovate-acuminate, often cordate at the base; or the basal and intermediate of each side combined into one deeply bilobed leaflet; the terminal leaflet of the quinate leaves is sometimes although rarely deeply lobed at the base; petioles (which are furrowed above) and midribs with strong large-based deflexed prickles beneath; stipules lanceolate or narrow.

Flowering shoot from fuscous scales, slightly wavy, slightly hairy, or with a fine coat of felt, a few minute aciculi and setæ. *Prickles* small, deflexed from a long base. *Leaves* ternate. *Leaflets* narrowed to the base; basal very unequal-sided; terminal broadly oval, acuminate, with a cordate base. *Panicle* rather long, open, with longish axillary racemose-corymbose branches, one or two of which are often very long and leafy and form secondary panicles; top formed of clusters of nearly simple peduncles in irregular corymbs; rachis and peduncles felted, setose, often hairy, with short slender declining prickles. *Sepals* greenish, broadly ovate-acuminate, with a slender (usually very short) point, felted, setose, rarely slightly aciculate, clasping the blue-black fruit. *Petals* contiguous, broad, wavy, usually jagged, nearly white. *Filaments* white. *Anthers* greenish. *Styles* pale green or slightly flesh-coloured, especially at the base. *Primordial fruit-stalk* about as long as the sepals. The panicle and sepals are only slightly armed; the prickles short; the setæ very short and nearly equal; the aciculi very few.

This seems to be the *R. althæifolius* (Host), and agrees in nearly all respects with Trattinnick's description. The quantity of hair beneath the leaves appears variable, and perhaps what is called felt in my description might more correctly be considered as a dense mass of interlacing hairs all seated upon the veins, for apparently the intervening spaces are naked. There is a good representation of the leaves of this plant is *Rubi Germanici* (tab. 45. A. f. 2), but the whitely-felted underside is rarely found: both the shoots and the panicle are figured as far more prickly than is usually the case with those parts of our plant.

R. nemorosus (Hayne) is not the same as my *R. althæifolius*, but, judging from his plate (*Arzneyg*, iii. t. 10), is nearly related to *R. corylifolius* γ *purpureus*. It has slightly stalked and incumbent basal leaflets, patent sepals, broad and almost triangular-ovate blunt pinkish slightly clawed petals, yellowish anthers, pinkish styles, black fruit, an oblong stalked torus, and few straight slender prickles upon the peduncles and rachis. Sonder considers it as identical with *R. pallidus* (Weihe), but their similarity in not apparent to me.

Sonder believes that the *R. nemorosus* (Arrh.) and *R. Wahlbergii* (Arrh.) are forms of *R. corylifolius* (Sm.); and as far as my information extends, I hold similar views concerning them. My *R. nemorosus a glabratus* is probably identical with *R. corylifolius a sublustris;* the authentic *R. Wahlbergii* from Sweden is apparently the *var.* γ *purpureus;* to *var.* β *conjungens* or to *var.* γ *purpureus* my *R. nemoro-*

sus β pilosus appears to be referable. Dr Bell Salter confounded the *R. Grabowskii* with his *R. Wahlbergii*, and has therefore confused their synonyms, localities and descriptions. Metsch quotes the *R. bifrons* (Vest.) to this plant. It is possible, but scarcely probable, that he is correct.

The *R. nemorosus δ ferox* (Leight.) is described below as a new species under the name of *R. tuberculatus*.

Such confusion exists concerning *R. nemorosus* that it is not easy to determine its synonymy: a matter of little consequence to us, because all the plants so-called in Britain justly claim other names. A difference is especially to be noticed in the accounts given of the calyx. Some authors state that it is reflexed after the flower has faded; and others, that it clasps the fruit. Arrhenius is ambiguous in this part of his description, for he only says "sepala sub anthesi reflexa;" Weihe and Nees say "calyces fructui adpressi;" Godron "reflechis à la maturite" and "fructu maturescente patula." In the second edition of Bluff and Fingerhuth the words are "calyce fructifero erecto." I do not know the character of the prickles of the true *R. nemorosus*, for my few specimens named *R. nemorosus* and *R. dumetorum* are not conclusive. In Fries's *Summa* we are told that *R. corylifolius* of the *Svensk Botanik* (t. 187) and *Herbarium Normale* (ix. 50) is *R. nemorosus var. ferox*. In my copy of the *Herb. Norm.* no barren stem is given and the flowering shoot is very like some states of our *R. corylifolius:* the former reference directs us to a plate on which a plant is represented having both of its shoots thickly covered with aciculiform prickles directed upwards (a state of them never found in nature); it looks like a bad figure of *R. corylifolius a sublustris*, rather than of *R. nemorosus*. The *R. nemorosus* of Arrhenius, as exhibited in the *Herb. Norm.* (vi. 47) is probably my *R. corylifolius β conjungens*. *R. dumetorum*, as illustrated by a specimen from "Apenrade Slesvigiæ"

sent to me by Mr Lange of Copenhagen, is *R. Balfourianus;*
one named *R. nemorosus* by the same skilful botanist, from
Flensburg in Slesvig, is much like some of my specimens of
R. corylifolius. It thus seems probable that the typical *R.
nemorosus* of Arrhenius and the *R. dumetorum* of Weihe are
really not separable from *R. corylifolius* (Sm.). Their *var.
ferox* is perhaps *R. diversifolius* (Lindl.) but, as already remarked when discussing that species, the figure in *Rubi
Germanici* represents a plant which is far more prickly on
the peduncles and petioles of its flowering shoot than any
R. diversifolius which I have seen ; also a slight bloom is
represented as existing on its stem.

Dr Salter's *R. nemorosus* is unintelligible to me. The
specimen in his *Herbarium* is very curious. Its barren
stem much resembles that of *R. cæsius β tenuis*, but is said
by him to have quinate leaves with the lower leaflets incumbent. Its panicle is open, exceedingly prickly above, the
sepals are large, long, and loosely clasp the fruit. Judging
from the only specimen which I have seen, I incline to refer
it to *R. althæifolius.*

If led by first appearances we might think that the *R.
althæifolius* is identical with the *R. Mougeoti* (Billot); but
that bramble has few strong and deflexed prickles on its
angular and furrowed stem, and its fruit-sepals are reflexed
and without glands or aciculi. A specimen of it is given in
the *Fl. Gal. et Germ. exsic.* No. 541, and it is described by
Billot in Schultz, *Archives*, 166 (1850).

The specimen obtained by Borrer from the authentic
bush in the Horticultural Society's Garden of *R. dumetorum*
(Lindl.) is certainly this species. It is probably the *R.
dumetorum* of both editions of Lindley's *Synopsis*, but is
identified with certainty as that of the second.

It is not with satisfaction that I find it necessary to
adopt new names, but the impossibility of avoiding it will

probably be admitted by most botanists who do not remove
the necessity by greatly reducing the number of recognized
species. Although the present name is not actually new, it
is so in effect, having fallen totally out of notice and never
even been quoted as a synonym in Britain until used by me.
As the plant agrees excellently with the original description its use can hardly cause any confusion. Nevertheless
there is a possibility that our plant may not be exactly that
of Host, for few brambles are absolutely identical in distant
parts of Europe, and Baker on the authority of a specimen,
names it *R. ligerinus* (Genev.).

A form of what seems to be this species from N. Yorkshire is named *R. degener* (Müll.) by M. Genevier. It has
no felt on the underside of its leaves but scarcely differs in
other respects. Other specimens from N. Yorkshire which
I refer confidently to *R. althæifolius* are named *R. degener*
by M. Genevier. Another is referred as a form to the *R.
Mougeoti* noticed above, but differs from that plant in the
manner there stated. It is also said by him to be the *R.
acerosus* Müll., but the specimens he sent to Mr Baker as
R. acerosus are *R. corylifolius a sublustris*.

Habitat.—Hedges. July and August.

Area.—1 2 3 4 5 . . . 9 10 11 12.

Localities.—i. Kew Stoke, *N. Som.*—ii. Bembridge,
Isle of Wight; Henfield and Steyning, *W. Suss.*—iii. Goldings and Mangrove Lane, *Herts.;* Pinner and Harrow,
Middl. (Hind!); Lea Bridge road, *S. Essex* (E. Forster!)—
iv. Eversden, Comberton, Balsars Hill and other places,
Cambr.—v. Henwick, *Worc.;* between the Brick-kiln pool
and Wilton road, Ross, *Heref.* (Purchas); Ham Lane, Cheltenham, *E. Glouc.* (Notcutt).—ix. Frodsham, *Chesh.*—x.
Thirsk, *N. E. York.*—xi. *Durham.*—xii. Douglas, *Isle of Man.*

42. R. tuberculatus Bab.

R. caule arcuato-prostrato subangulato sparsim
brevi-piloso et -setoso, *aculeis* crebris inæqualibus tenuibus *e basi oblonga tuberculiforme* patentibus, foliis ternatis vel quinatis, foliolis subduplicato-dentatis subtus
in venis pilosis utrinque viridibus inferioribus foliorum
ternatorum bilobatis *infimis foliorum quinatorum* subsessilibus *intermediis incumbentibus,* foliolo terminali
rotundo-cordato subcuspidato, paniculæ foliosæ ramis
axillaribus racemosis apice corymbosa aculeis a medio
usque ad apicem paniculæ et pedunculorum tenuibus quam reliquis longioribus, sepalis ovato-acuminatis
aciculatis setosis fructui laxe adpressis.

R. tuberculatus Bab.! Fl. of Camb. 306 (1860); Man.
ed. 6. 119. Syme's Eng. Bot. iii. 194.

R. nemorosus δ *ferox* Leight.! Shrop. Rubi (sp.). Bab.!
Man. ed. 3. 104; ed. 4. 107.

˙ *R. dumetorum* Blox.! Fasc. Rub. (sp.).

R. scabrosus Müll. Mon. 196.

Stem very bluntly angular, with scattered short hairs
and setæ, reddish. *Prickles* many, conical, slender, rising
rather abruptly from large oval depressed tubercles which
are often purplish, patent. *Leaves* ternate or rarely quinate.
Leaflets irregularly and somewhat doubly dentate-serrate,
dull green, rugose and pilose above, pale green, hairy on the
veins, and very finely or slightly felted beneath; basal of the
ternate leaves bilobed, lower lobe usually rounded and blunt

but sometimes acute, upper broad roundly oval cuspidate; terminal roundly or rather quadrangularly cordate, subcuspidate; basal of the quinate leaves obovate, usually blunt, subsessile or shortly stalked overlapping the unequal-sided obovate cuspidate intermediate leaflets; petioles (which are furrowed above) and midribs with declining prickles beneath; stipules narrowly lanceolate or rarely linear.

Flowering shoot from ashy scales, slightly angular, covered with a thin coat of fascicled crisped hairs many short setæ and aciculi. *Prickles* unequal, slender, declining, from a long compressed base. *Leaves* ternate. *Leaflets* serrate, doubly or lobate-serrate towards the tip, pilose above, hairy on the veins beneath, green on both sides; basal bilobed, sessile; terminal obovate, narrowed to a slightly truncate base. *Panicle* rather short, leafy; axillary branches racemose, few-flowered, ascending; ultra-axillary top corymbose; rachis and peduncles finely felted, hairy, with long unequal setæ and aciculi, and long slender declining prickles which are longest at the top of the panicle and of the peduncles where they are abundant. *Sepals* ovate-acuminate, long-pointed, felted, with many hairs setæ and aciculi, clasping the fruit. *Petals* roundly obovate, jagged at the end, pinkish. *Stamens* and *styles* greenish-yellow. *Nut* obovate-oblong, the point of attachment and style lateral.

We are indebted to Mr Leighton for pointing out the existence and characteristics of this plant, of which he kindly supplied me with an abundance of specimens, and also with valuable remarks concerning it. It cannot be the *R. nemorosus* γ *ferox* of Arrhenius, nor the similarly named variety of *R. dumetorum* of Weihe, nor the latter botanist's *R. ferox*; for that plant has not the tubercular-based prickles which are characteristic of *R. tuberculatus* and is much more prickly on the flowering shoot; in the words of Weihe and

Nees, "pedunculi et calyces aculeis glandulis pilisque valde horrentes;" and a similar description is given by them of the barren stem. The armature of the stem of our plant is totally different; the short hairs and setæ, although tolerably abundant, being inconspicuous. On the flowering shoot the petioles are distantly furnished with prickles and have few aciculi or setæ; the upper part of the rachis, and of the peduncles, bears an abundance of long straight or slightly deflexed slender prickles which much exceed in length the few aciculi and (often) rather plentiful but unequal setæ: but the number of the setæ is very variable, even upon the same bush. The sepals also are much less strongly armed than those of *R. ferox* (Weihe). It should be added that I have derived all my knowledge of the *R. ferox* from the imperfect specimen contained in the *Herbarium Normale* of Fries (ix. 50), the plate in the *Rubi Germanici*, and the description in Arrhenius's *Monograph*. I believe it is referable to *R. diversifolius* (Lindl.).

Plants belonging apparently to this species are tolerably abundant in the county of Cambridge. They have more interlacing hairs on the barren stem than are found on Leighton's specimens. The stipules of the barren stem are variable in form being sometimes lanceolate and at others very narrow.

M. Questier has sent this plant as *R. dumetorum var. ferox*. I have already endeavoured to show that it cannot bear that name. It is clearly not the *R. ferox* of Boreau, nor do I find any description which will suit it in that author's elaborate account of the *Rubi* of central France.

It may be useful to add how all the forms of the supposed English *R. nemorosus* (Man. ed. 4. 107) are disposed of—a = *R. corylifolius* a; β = *R. corylifolius* γ; γ = *R. corylifolius* β; δ = *R. tuberculatus*.

Habitat.—Hedges. July, August.

Area.— . 2 3 4 5 . . 8 . 10 . 12 . . . . 21 . 23.

Localities.—ii. Henfield, *W. Suss.*—iii. Richmond, *Surr.* —iv. Caldecot, Wood Ditton, Wicken, and a few other places in *Cambr.;* Hunstanton, *W. Norf.*—v. Llanrumney, *Monm.;* near *Worcester* (Lees!); Red Hill near Shrewsbury, *Salop;* Michel Dean, *W. Glouc.*—viii. Twycross, *Leic.*—x. Thirsk, *N. E. York.*—xii. Alston, *Cumb.*

xxi. By the river at *Kilkenny.*—xxiii. New Grange, *Meath.*

43. R. cæsius Linn.

R. caule prostrato tereti pruinoso, aculeis parvis inæqualibus e basi longa compressa declinatis deflexisve, foliis ternatis rarissime pinnatis, *foliolis inæqualiter inciso- vel grosse-serratis*, terminali ovato rhombeo-ovato vel trilobo lateralibus subbilobatis subsessilibus, panicula subsimplici sæpe depauperata, sepalis ovatis acuminatis apice longa lineari fructui (cæsio-pruinoso) adpressis, petalis obovatis emarginatis, stylis virescentibus.

R. cæsius Linn. Fl. Suec. ed. 2. 172 (1755); Sp. Pl. 706. Sm. Eng. Fl. ii. 409; Eng. Bot. t. 826. Arrh. Mon. 50. Lees! in Steele 54; Malv. 50. Leight.! Fl. Shrop. 238. Blox. in Kirby 37. Bab.! Man. ed. 6. 119. Sond. Fl. Hamb. 285. Metsch in Linnæa xxviii. 107. Syme's Eng. Bot. iii. 195. t. 456.

R. cæsius α Borr. in Hook. ed. 2. 248; ed. 3. 251.

R. cæsius et *R. agrestis* Merc. in Reut. Fl. Genev. 262 and 263.

α *umbrosus;* caule tenuissimo, aculeis paucis parvis, foliolis planis utrinque sparsim pilosis lobato-serratis terminali rhomboideo-ovato acuminato basi rotundo.

R. cæsius α *umbrosus* Reichenb. excurs. 608. Arrh. Mon. 50.

R. cæsius α *agrestis* Bab.! Man. ed. 5. 109.

R. cæsius α *aquaticus et* β *agrestis* Rubi Germ. 102. t. 46. A.

R. cæsius ε *tenuis* Lees! in Steele 54.

R. cæsius Blox.! Fasc. (sp.).

R. ligerinus Genev.! Mem. Soc. Acad. Angers. viii. (1860).

Stem prostrate, round, glaucous-green, slender, with very small scattered declining or deflexed prickles from slightly enlarged bases; hairs, setæ and aciculi very few or wanting. *Leaves* ternate. *Leaflets* thin, flexible, flat, dull green, and pilose above, rather glaucous, and with hairs on the veins beneath, lobate-serrate; basal subsessile, ovate-acuminate or broadly lanceolate, with a large lobe on the lower side; terminal long-stalked, rhomboidal-ovate, acuminate; petioles (which are furrowed above) and midribs with few very small prickles beneath; stipules linear-lanceolate.

Flowering shoots from fuscous scales, resembling the stem. *Prickles* very few and very small. *Leaves* ternate. *Leaflets* rhomboidal-ovate, lobate-serrate, resembling those of the stem. *Panicle* small, leafy, with rather few short declining prickles, and a few setæ and aciculi; flowers in terminal axillary corymbs. *Peduncles* felted, setose, prickly. *Sepals* ovate, acuminate, with a rather long point, greenish with a narrow white margin, felted, slightly setose, clasping the fruit. *Petals* white, obovate, with two or three notches at the end, clawed. *Filaments* white. *Anthers* cream-coloured. *Styles* greenish. Sometimes the stamens and styles are yellowish. *Fruit* of few large black glaucous drupels.

The panicle of this plant is often nearly simple, and when otherwise the branches are rarely more than once divided; they are very variable in length, and are sometimes exceedingly long.

β *tenuis;* caule tenuissimo, aculeis crebris validis sed parvis subæqualibus deflexis, foliolis planis (?) utrinque pilosis vel subtus villosis duplicato-serratis terminali obovato vel cordato-obovato acuminato, fructu nigro "nec cæsio-pruinoso."

R. tenuis Bell Salt.! in A. N. H. xv. 305 (1845). Bab.!
Syn. 11; Man. ed. 2. 98.

R. cæsius β tenuis Bab.! Man. ed. 6. 119.

R. cæsius Leight.! Shrop. Rubi, 26 (sp.).

R. parvulus Genev.! in Mem. Soc. Acad. Angers. viii.
(1860).

This plant differs from *var. α* in the following respects.
The prickles on the stem are many, small, stout, much deflexed, mostly equal, from considerably enlarged bases. A
very few aciculi and short setæ are sometimes found. Dr
Bell Salter mentions quinate leaves, but the only approach
to them which I have seen is found on a specimen sent by
Mr Bloxam, where one leaf has four leaflets of which the
fourth is very deeply lobed. The leaflets are rather doubly
than lobate-serrate; sometimes the underside has such an
abundance of hairs as to seem felted, but such is also the
case (although rarely) in var. *α*; the terminal leaflet is always, but sometimes only slightly, narrowed below, and
occasionally has a cordate base. The flowering shoot and
flowers do not seem to differ, but the fruit is said to want
the bloom usually found upon that of the other varieties.

This has been supposed to be the *var. agrestis* of Weihe
and Nees, and rather strong states of it are often so-named;
but apparently their plant is only a slight variation of their
var. umbrosus, having stronger stems, more prickles, leaflets
which are rugose above and densely hairy beneath, and a
rather rounder terminal leaflet.

Dr Bell Salter's *var. ferox* of his *R. tenuis* (A. N. H. xv.
305) agrees with the usual state of the plant in all respects,
except in having many exceedingly strong, short compressed,
deflexed prickles upon both its shoots. It has only been
found in one place, viz. near the farm at Apse Down, Isle of
Wight. It is the *R. cæsius δ ferox* (Bell Salt.!) in the *Bot.
Gaz.* (ii. 130) and *Fl. Vect.* (160); but not the *R. nemorosus*

c. ferox (Arrh.), nor the *R. dumetorum* δ *ferox* of the *Rubi Germanici.*

Intermediate forms between the varieties *ferox* and *tenuis* are not unfrequent.

γ *ulmifolius;* caule tenui purpurascente, aculeis crebris parvis deflexis declinatisve, aciculis setis pilisque paucis brevibus, foliolis rugulosis lobato-serratis subtus in venis pilosis vel hirtis vel cinereo-subtomentosis terminali rotundo-cordato acuminato sæpe *trilobo* vel rarissime in tribus foliolis sessilibus diviso cum extimo ad basin attenuato.

R. ulmifolius "Presl. Del. Prag. 223"? (not Schott).

R. cæsius γ *ulmifolius* Bab.! Man. ed. 5. 110; ed. 6. 119.

R. cæsius β *pseudo-cæsius* Weihe in Boenn. Prod. Fl. Monast. 151. Rubi Germ. 103. t. 46. B. f. 1. Metsch in Linnæa xxviii. 109.

R. cæsius β *arvensis* Wallr. Sched. 228. DC. Prod. ii. 558.

R. cæsius β *rugosus* Lees! in Steele 54.

R. pseudo-cæsius Lej. Rev. de la Fl. de Spa. 101.

R. Idæo-cæsius Wirtg.! Herb. Rub. No. 116 (sp.).

Stem slender but often thicker than in the preceding varieties, purplish or purple, but also slightly glaucous, round, with many small declining (or on the autumnal shoots deflexed) prickles; hairs, setæ and aciculi very few and short. *Leaves* ternate. *Leaflets* thin, flexible, nearly flat, slightly rugose, dull green and pilose above, hairy on the veins beneath (sometimes so densely as to seem felted), lobate-serrate; basal ovate but very unequal-sided, or deeply bilobed, or rarely the lobes separate and form an oblong sessile basal and a slightly stalked unequal-based broadly ovate intermediate leaflet; terminal longstalked, roundly cordate, shortly acuminate, usually three-lobed, the lateral lobes usually rounded at the end and more or less deeply separated from the terminal lobe, or divided into three sessile leaflets of which the terminal is usually narrowed to its base; petioles (which are furrowed above) and midribs with very small prickles beneath; stipules linear-lanceolate.

Flowering shoot from fuscous scales, slightly glaucous especially near to the base. *Prickles* small, many, declining. *Leaves* ternate. *Leaflets* lobate-serrate, broad; lower very unequal-based; terminal slightly narrowed to the base. *Panicle* loose, rather corymbose; peduncles finely felted, setose. *Sepals* slightly setose, with a few aciculi at the base. *Petals, stamens,* and *styles* as in *var. a.*

This is often a very much larger plant than either of the preceding varieties. Its stems are thick and strong, although quite prostrate unless supported. Its leaves are very broad and often exceedingly hairy beneath in such a manner that (although all the hairs spring from the veins) the surface is quite hidden: but sometimes the hairs are few and to be detected with difficulty. It is usually so different from the ordinary form of *R. cæsius* that many persons have considered it as a distinct species; and, as will be seen above, it has even been published as such by botanists of repute. If distinct it cannot bear the name of *R. ulmifolius* which belongs to a plant found at Gibraltar. I am convinced that it is a form of *R. cæsius.*

δ *intermedius;* caule crassiore viridi-purpurascente, aculeis crebris tenuibus subpatentibus valdè inæqualibus, aciculis setisque paucis brevissimis validis, *foliolis* lobato-serratis subtus in venis pilosis *terminali triangulari-cordato* acuminato trilobo vel tripartito *vel in tribus foliolis sessilibus diviso,* cum extimo ad basin attenuato.

R. cæsius δ *intermedius* Bab.! Man. ed. 5. 110; ed 6. 120.
R. dumetorum γ *bifrons* Lees! in Steele 54.

This variety seems to connect the var. *ulmifolius* with the var. *pseudo-Idæus.* The stem is stronger, the prickles are larger, the aciculi and setæ shorter and stronger, the leaflets different in shape from those parts in the var. *ulmifolius.* It differs from var. *pseudo-Idæus* by its stem being less pruinose, the prickles, aciculi and setæ stronger, the terminal leaflet of the pinnate leaves different in shape, sessile, and narrowed (not rounded) to its base. Most of the leaves are quinate.

ε *pseudo-Idæus;* caule crassiore viridi et eximie pruinoso, aculeis tenuibus violaceis subpatentibus, aciculis et setis brevissimis paucis, *foliis* ternatis vel *quinato-pinnatis,* foliolis duplicato-serratis subtus cinereo-tomentosis, lateralibus omnibus sessilibus *terminali petiolato rotundo-cordato.*

R. cæsius δ *pseudo-Idæus* Weihe in Boenn. Prod. Fl. Monast. 151. Rubi Germ. 104. t. 46 B. f. 2. Bab.! Man. ed. 5. 110; ed. 6. 120. Webb and Colem.! Fl. Hertf. 85.
R. pseudo-Idæus Lej. Rev. 102.

Stem prostrate, round, green, glaucous, very finely felted, with a few short setæ. *Prickles* small, patent, many, violet-coloured, as are also their oblong very slightly elevated bases. *Leaves* quinate-pinnate. *Leaflets* thin, flexible, flat, dull green and slightly pilose above, paler and finely felted but nearly or quite without hairs beneath, slightly lobate-serrate;

lower sessile, ovate, acuminate, strongly lobed on the lower side, or each divided into two leaflets, of which one is sessile and ovate and the other very shortly stalked and broadly ovate ; upper pair separated considerably from the basal, sessile, ovate, acute, unequal-sided; terminal stalked, roundly cordate, acuminate; petioles and midribs with small straight slender prickles beneath; petioles slightly furrowed above, rather more furrowed beyond the lowest leaflets ; stipules linear-lanceolate.

Flowering shoot rather slender. *Prickles* straight, declining, increasing in number towards the panicle. *Panicle* long, leafy, with very many slender declining prickles on its upper part and peduncles, and few or no setæ, felted. *Flowers* in small axillary and terminal, few-flowered corymbs. *Sepals* ovate-acuminate, with a long slender point, green, aciculate, setose, felted, clasping the fruit.

I have only seen one specimen of this variety, which was given to me by my lamented friend the Rev. W. H. Coleman, its discoverer at Hunsdon in Hertfordshire.

I have received specimens bearing the name of *R. pseudo-Idæus* or of *R. cæsius β pseudo-Idæus* from Dr Lindeberg (in Bahusia, raro), and Prof. Lange (Apenrade, Slesvigiæ), which almost exactly agree with the plate in *Rubi Germanici.* They and the plant represented on that plate differ from our plant from Hunsdon by having the terminal leaflet of the pinnate leaves narrowed to the base, whereas on our plant from Hertfordshire it is much the broadest and cordate at the base. Dr Metsch (Linnæa, xxviii. 105) refers the plants of Weihe and of Lejeune to *R. Idæus* as *β canescens*, but I cannot agree with him. The *R. pseudo-Idæus* of Müller is, as he suspected, the *R. suberectus* of Anderson.

Mr Lees has sent a plant with the name of *R. cæsius v. pseudo-Idæus* from woods near Worcester, which is not exactly my plant nor that of the *Rubi Germ.* It has quinate-pinnate leaves which are not felted beneath, nor is the terminal leaflet rounded, but is much narrowed below, and in one case very nearly sessile. It seems to connect *r. pseudo-Idæus* with *v. ulmifolius*, and is an additional proof, if proof was wanted, of the plants here included under *R. cæsius* forming only one species.

ζ *hispidus;* caule tenui viridi, aculeis brevibus multis aciculiformibus inæqualibus, setis multis brevibus rigidis, foliis ternatis, foliolis lobato serratis subtus in venis pilosis lateralibus retrorsum bilobatis terminali obovato acuminato basi subcordato, pedunculis sepalisque valde setosis et tomentosis vix hirtis.

R. cæsius ϵ *hispidus* Rubi Germ. 104. t. 46 C. f. 1. Bab.! Man. ed. 5. 111; ed. 6. 120.

R. serpens Godr. in Fl. de Fr. i. 538; Fl. Lorr. ed. 2. i. 231. Bor. Fl. Centre, ed. 3. 187. Billot! Fl. Gall. et Germ. exsic. No. 762 (sp).

Stem slender, green, glaucous, glabrous; prickles, aciculi and setæ small, many, unequal; prickles subulate, patent or subpatent, straight, from an enlarged base, slightly deflexed on the autumnal shoots. *Leaves* ternate. *Leaflets* thin, flexible, flat, dull green and pilose above, paler and hairy beneath, lobate-serrate; lower subsessile, oblong, acute, with a large lobe on the lower side; terminal rhomboidal or ovate, slightly cordate at the base; petioles (which are furrowed above) and midribs with small nearly straight prickles and aciculi beneath; stipules leaflike, stalked.

Flowering shoot not so distinct from the stem as is usual, often springing from it during the first summer of its growth, and similar to it in form, clothing and foliage. *Panicle*

very irregular and lax, wavy, leafy, with many short red
setæ. *Flowers* in irregular terminal and axillary corymbs.
Peduncles very setose, finely felted. *Sepals* ovate-acuminate,
with a long slender point, pale green, felted, setose, as well
as the peduncles, scarcely at all hairy, spreading, clasping
the glaucous fruit, which consists of a few large drupels.
Petals white, ovate, notched, shortly clawed. *Filaments*
white. *Anthers* cream-coloured. *Styles* greenish. *Primordial fruitstalk* about as long as the sepals. After the
petals have fallen the sepals quite close over the drupels so
as to bring their flattened points together: in which state
they closely resemble the buds, scarcely differing except in
size and the length of the points: the swelling of the fruit
causes them again to open.

This variety seems to be or to include the *R. serpens*
(Godr.), with which two of my specimens almost exactly
agree. One of these, from Fakenham in Norfolk, has a
slight trace of bloom on its stem and straight prickles on its
petioles; it also has nearly naked unarmed sepals: the other
(from Buildwas in Shropshire), which is derived from the
Herb. Leighton, agrees better in these respects with *R.
serpens*. Both have much more numerous drupels in each
fruit than is usual in *R. cæsius*, which has rarely more than
1—3 large ones. Godron states that the fruit-calyx of *R.
serpens* is reflexed, which is not the case in our plant. The
specimen of *R. serpens* in Billot's *Fl. Gall. et Germ.* agrees
very well with the present variety.

Dr Mercier divides this species into *R. cæsius* and
R. agrestis by the "impari-cuneiform" terminal leaflet of the
former and the "impari-cordate" leaflet of the latter. I cannot find much other difference between his plants, and consider that character not even of use as separating varieties.

Mr Syme states that he is unable to distribute the specimens which he has seen into my varieties. I have rarely

found much difficulty in doing so, but it is to be expected that many intermediate forms should occur.

Considered as a whole *R. cæsius* seems to be a tolerably well-marked species. It is variable, and its extremes differ considerably; but probably no botanist who has paid much attention to brambles will have any doubts concerning its forming only one species.

Habitat.—Fields, hedges and heaths. June, July.

Area.—1 2 3 4 5 6 7 8 . 10 11 . 13 14 . . . . 19 20 . . . 24 . 26 27 28 . 30.

It is probable that all the other provinces produce this plant. Mr Watson adds doubtfully 9, 12, and certainly 15, to the numerical list of those of the correctness of which I have had personal proof. It will be seen that we have no record of *R. cæsius* from the north of Scotland, and that several of the Irish Provinces are not as yet known to produce it. It is not thought necessary to give exact localities for so common a plant.

Sec. II. Rubi Herbacei.

Caules herbacei.   Folia ternata vel simplicia.   Stipulæ ovatæ, cum petiolum caulem amplectentes.   Flores umbellati vel subsolitarii.   Receptaculum planum.

i. *Saxatiles.* Caules flagelliformes.   Flores umbellati vel subsolitarii.   Acini magni, pauci, discreti.

44. R. saxatilis Linn.

R. caule tenui prostrato radicante inerme vel aciculis parvis paucis herbaceis distantibus exasperato, foliis ternatis, ramo florifero erecto corymbifero paucifloro, petalis lanceolatis calycem subæquantibus.

R. saxatilis Linn. Fl. Suec. ed. 2. 173 (1755); Sp. Pl. ed. 3. 708. Eng. Bot. t. 2233. Sm.! Eng. Fl. ii. 410. Bab.! Man. ed. 6. 120. Rubi Germ. 30. t. 9. Reichenb.! Fl. exsic. 787 (sp.). Metsch in Linnæa, xxviii. 102. Garke, Fl. Deutsch, ed. 3. 108. Lange, Danske Fl. 340. Wirtg.! Herb. Rub. No. 50 (sp.). Syme's Eng. Bot. iii. 159. t. 441. *Chamærubus saxatilis* Raii Syn. ed. 1. 94; ed. 3. 261.

Stem annual, almost herbaceous, very slender, angular, pilose, prostrate, rooting at the end; prickles none, or few very small and weak. *Leaves* ternate. *Leaflets* about equal, oblong-obovate, unequally coarsely or doubly serrate, pale green on both sides, usually hairy on the veins beneath; lateral nearly sessile, unequal-sided; stipules lanceolate, narrower towards the end of the stem, sometimes attached to the petiole alone, usually to the petiole and stem; petioles hispid, slightly channelled above, with a few distant slender prickles beneath.

Flowering shoot erect, springing from the rhizome, angular, hairy. Leaves like those of the stem. *Flowers* corymbose, at the top of the shoot. *Sepals* triangular-lanceolate, reflexed when in flower, afterwards adpressed to fruit. *Petals* white, erect, inconspicuous, narrow. Fruit of few (3 or 4) roundish, fleshy, red drupels, which are quite distinct and fall off singly.

Mr Syme states that the sepals are "reflexed in fruit," but that is not my experience nor that of Arrhenius. Probably there is a slip of the pen here.

The rhizome is strong, woody, usually subterranean. The stems are not more than annual except a very short piece of their base, from which the stems and shoots of the succeeding year are thrown out.

Habitat.—Rocky places in woods in hilly districts, and on mountains.

Area.— .... 5 . 7 . 9 10 11 12 . 14 15 16 . 18 19 .... 24 .... 29 30.

Localities.—v. Queen's wood near Prestbury, *E. Glouc.* —vii. Wrexham, *Denb.;* above Llyn y Nadroedd, Snowdon, *Caern.*—ix. Kirkby Londsdale, *N. Lanc.*—x. Roch Abbey woods, *S. W. York;* Hawnby, *N. E. York;* Round How near Richmond, *N. W. York.*—xi. Castle Eden Dene and Heaton Dene (Winch), *Durh.;* Deyne near Hexham, also Wallow Crag (Robertson!) *Northumb.*—xii. Gilsland, *Cumb.*

xiv. Blackburn-rig, Dene, and elsewhere, *Berw.;* Roslin, *Edinb.*—xv. By the river Don at Aberdeen, Ben na Bourd, and Linn of Corrymulrie, *S. Aberd.;* Clova, *Forf.;* Craighall woods, *E. Perth;* Ben Lawers, *W. Perth* (Balfour); Glen Lochay, *Mid Perth* (E. Forster).—xvi. Coulin hills, Skye, *N. Ebudes.*—xviii. Hoy, *Orkney;* Euness, Unst, *Shetland.*

xix. Shores of the Lakes of Killarney, *S. Kerry.*—xxiv. Castle Taylor, *E. Galw.* (A. G. More!)—xxix. Ards and elsewhere in *Donegal* (E. Murphy in Loud. Mag. Nat. Hist. i. 437).—xxx. Glen Ariffe and upper part of Colin Glen, *Antrim.*—Common in the glens of the north, west and south of Ireland (D. Moore).

ii. *Arctici.* Caules steriles nulli. Rhizomata subterranea longa. Flores terminales, subsolitarii. Acini in baccam compositam congesti.

45. R. Chamæmorus Linn.

R. caule erecto inermi unifloro, floribus dioicis, foliis simplicibus lobatis plicatis.

R. Chamæmorus Linn. Fl. Suec. ed. 2. 174 (1755); Sp. Pl. 708. Eng. Bot. t. 716. Sm. Eng. Fl. ii. 412. Reichen.! Fl. Germ. exsic. 2174 (sp.). Bab.! Man. ed. 6. 120. Rubi Germ. 113. t. 49. Arrh. Mon. 57. Lange, Danske Flora, 340. Syme's Eng. Bot. iii. 158. t. 440.

Stems subterranean, creeping extensively. *Flowering shoots* erect, 3—8 inches high, unarmed, with 3 or 4 distant ovate scales below, leafy above. *Leaves* 2 or 3, alternate, stalked, reniform, with 5 blunt unequally dentate lobes, . plicate, with hairs and stalked glands on the veins beneath or on both sides. *Stipules* ovate, attached to the stem. *Flower* terminal, large, dioecious. *Sepals* ovate. *Petals* obovate, large, white. *Fruit* very large, red, but becoming orange-yellow when ripe, of few large drupels.

Habitat.—Turf-bogs on mountains.

Area.— . . . . . . 7 8 . 10 11 . . . 15 . . . . 28.

Localities.—vii. Cader Fronwyn, *Flint.*—viii. Edal Cross, *Derby.*—x. Cronkley Fell, *N. W. York.*—xi. *Durham* (Herb. Borr. !); Cheviot, *N. North.*

xv. Ben Lawers, *Mid Perth;* Clova and Glen Isla, *Forf.;* Loch na Gar, *S. Aberd.*

xxviii. Stranagalvally Mountains (Mackay), Glen Garro Mountains (E. Murphy in Loud. Mag. i. 437), *Tyrone.*

Mr Watson adds Provinces 9 . . 12 13 14 . 16.

R. arcticus Linn.

R. caule erecto inermi subunifloro, foliis ternatis, petalis obovatis calyce duplo longioribus, staminibus conniventibus, acinis multis.

R. arcticus Linn. Fl. Suec. ed. 2. 173 (1755); Sp. Pl. 708. Eng. Bot. t. 1585. Sm. Eng. Fl. ii. 411. Rubi Germ. 111. t. 48. Arrh. Mon. 55. Fries! Herb. Norm. iii. 44 et xii. 53. Bab.! Man. ed. 5. 120.

Stem subterranean. *Flowering shoots* erect, 4-6 inches high, unarmed, with 3 or 4 distant ovate scales below, leafy above. *Leaves* ternate. *Leaflets* nearly equal, broadly ovate, crenate. *Flower* usually solitary, terminal; or one or two additional flowers opposite to the upper leaves. *Sepals* oblong-lanceolate. *Petals* rosecoloured, obovate, variable in number. *Fruit* of many cohering drupels.

Habitat.—Turfy bogs on mountains.

Area.— . . . . 15 16.

Localities.—Said to have been found in the Isle of Mull by the Rev. Dr Walker, and on Ben y Glo in Perthshire by Mr R. Cotton. Probably some mistake has occurred in each case, although there is a specimen ticketed as from the latter place in Sowerby's Herbarium at the British Museum.

POSTSCRIPT.

Dr D. Moore has sent to me a specimen of *R. laciniatus* (Willd.), which he received from Hollypark in the county of Dublin; but he has never seen the plant growing wild, and knows nothing of its history. Apparently it is of garden origin, in common with all those which bear the name of *R. laciniatus:* for Seringe (*DC. Prod.* ii. 558) says "Patria ignota;" Willdenow describes it from the Berlin Garden (*Hort. Berol. fol. et tab.* 89); and it is figured by Watson (*Dendr. Brit.* t. 69) from a nursery garden near London. It was figured as a garden plant by Plukenett in 1691 (*Phytogr.* t. 108. f. 4). It is not unfrequently found in the gardens of the curious. It is not a state of *R. thyrsoideus,* as supposed by Willdenow and Weihe and Nees, for it is one of the *Sylvatici;* it has no apparent relationship to the *R. corylifolius* (Sm.), as thought by Wallroth. As it is not known to be a native of Britain, I am not called upon to determine its true place in the Genus nor to describe it here. I have also received specimens from near Truro in Cornwall and Grasmere in Westmoreland. The former is less certainly the *R. laciniatus* than the latter, which is exactly the garden plant.

INDEX.

The names printed in small capitals are those adopted for species or sections: the others are synonyms or species noticed. The numbers refer to the pages.

26

CAMBRIDGE: PRINTED BY C. J. CLAY, M.A. AT THE UNIVERSITY PRESS.